Principles of
Vehicle Body Repair

by the same author

Fundamentals of Vehicle Bodywork

Principles and Practice of Vehicle Body Repair

J. Fairbrother, TEng., LMIBE, MIBCAM

Chief Examiner, City and Guilds of London Institute,
 Vehicle Body Engineering Technicians
Senior Lecturer in Vehicle Body Crafts Studies and Body Engineering,
 Department of Automobile Engineering,
 Willesden College of Technology

STANLEY THORNES (PUBLISHERS) LTD

© J. Fairbrother 1983

All rights reserved. No part of this publication may be reproduced or transmitted in any form or by any means, electronic or mechanical, including photocopy, recording, or any information storage and retrieval system, without permission in writing from the publisher or under licence from the Copyright Licensing Agency Limited. Further details of such licences (for reprographic reproduction) may be obtained from the Copyright Licensing Agency Limited, of 33-4 Alfred Place, London WC1E 7DP.

Originally published in 1983 by Hutchinson Education
Reprinted 1985, 1988

Reprinted 1990 by
Stanley Thornes (Publishers) Ltd
Old Station Drive
Leckhampton
CHELTENHAM GL53 0DN

British Library Cataloguing in Publication Data
Fairbrother, J.
    Principles and practice of vehicle body repair.
    1. Motor vehicles — Bodies — Repairing
    II. Title
629.2'6   TL255

ISBN 0 7487 0280 6

Set in Times

Printed and bound in Great Britain at
The Bath Press, Avon

Contents

Preface   7
Acknowledgements   8

1 Introduction to body repair   9

Strength of metals – Insurance – Repair or replacement – Estimating – Requirements of body repair shop – Powered tools – Health and safety – Repairs to glass-fibre reinforced plastics bodywork

2 Refinisher protection and safety   31

Safety signs – Legal requirements for materials and employees – General regulations for siting and workshop facility – Environmental health and safety for refinish operators

3 Hand tools and panel beating techniques   39

Damage assessment – Panel preparation – Beaters and beating – Spoons and their applications – Specialized tools – Panel finishing – Metal finishing – Care of tools

4 Welding processes   54

Historical background – Oxy-acetylene welding – Oxygen cutting – Resistance spot welding – Metal arc gas-shielded (MAGS) welding – Tungsten arc gas-shielded (TAGS) welding – Metallic arc welding

5 Collision damage repair equipment   78

Development – Porto-Power – The Dozer technique – Operation of conventional Porto-Power – Operation of Dozers – Body bay systems (Flexi-Force) – Repair bench and measuring systems – General repair techniques

6 Refinishing facility and equipment   102

The refinish facility – Compressed air supply and equipment – Preparation – Spray booths – Spray guns – Other equipment – Basic refinishing rules

7 Spray painting principles   124

Flow cup and viscosity – Spray gun adjustments – Spraying technique – Spray gun maintenance – Spray gun problems – Motion study – spraying procedure – Paint heating – Electrostatic hand gun spraying

8 Paint materials and refinish processes   139

Technical terms – Paint characteristics – Undercoats – Finishes – Thinners – Metallics – Refinish processes and operational sequences – Special processes

9 Refinishing systems   158

Identification of original finish – Cellulose colour system – Acrylic colour system – Acrylic modified synthetic (2K, two-pack) colour system – Synthetic colour system – Clear coat metallics

10 Paint defects: cause, prevention and repair   170

11 Multiple-choice questions   187

Answers to multiple-choice questions   198
Index   199

Preface

This book will develop understanding and general appreciation of the craft principles and job knowledge required of the light vehicle body repairer and paint refinisher. It outlines techniques, equipment, materials, processes, procedure and safety.

Although primarily intended for apprentices receiving training in light vehicle body repair and paint refinishing, and for students following the City and Guilds of London Institute courses 385 Vehicle Body Craft Studies and TEC Vehicle Body Engineering courses, *Principles and Practice of Vehicle Body Repair* will be of interest to all students following motor vehicle courses, motor car owners and the young people in schools looking into the activities of the motor industry.

The ever increasing development of modern technology on motor vehicles requires the advancement of operator knowledge, and body repair work is no exception. Recent years have seen many improvements to both products and processes. As repair and paint refinishing are labour-intensive operations it is essential that this advancement is reflected in further education and training needs.

The advantages obtained from the use of modern techniques by highly skilled staff cannot be overstressed. Many vehicles require rework owing to inadequate or incorrect repair and refinish processes and many fail to make the original manufacturing finish owing to incompatibility of materials used, etc. These and many other problems are often the result of inadequate repair/refinisher knowledge and can be extremely costly in terms of lost profit and customer satisfaction.

Craftsmanship is predominantly the result of years of experience and practical application, but this alone will not suffice to ensure that the full benefits from modern technology are realized.

After a substantial and vigorous yet rewarding apprenticeship as a wheelwright and carriage builder and motor body builder and repairer, after industrial experience as a builder, repairer and finisher of motor vehicles, and after over twenty years of lecturing in vehicle body craft studies and body engineering, I would like to share my experience and cumulation of teaching notes so that students of motor vehicle work may benefit from this work.

J. Fairbrother

Acknowledgements

I wish to thank the following firms and/or their representatives who have co-operated with me by permitting the use of technical literature and guidelines on recommended processes, procedure and safety precautions. I am also grateful to my wife Sylvia and my daughter Helen for their help, patience and encouragement during the preparation of this book.

ARO Machinery Co. Ltd
Ault and Wiborg Paints Ltd
Berger Vehicle Refinishes
Blackhawk Automotive Ltd
British Leyland Service
British Oxygen Company Ltd
Car Bench Ltd
Celette/Churchill Ltd
Flexiforce Automotive Ltd
Ford Motor Company Ltd
Imperial Chemical Industries Ltd
Kango Wolf Power Tools Ltd
Samefa (UK) Ltd
Sykes-Pickavant Ltd
The De Vilbiss Company Ltd
The Valentine Varnish and Lacquer Co. Ltd

1 Introduction to body repair

Strength of metals

The ultimate in the design of a car body is to give maximum strength with minimum of weight, and it is essential that the light vehicle body repairer has at least an elementary knowledge of the methods of attaining this combination.

In the early days of the automobile industry, bodies were constructed mainly of wood by craftsmen well versed in the building of horse-drawn carriages and the vehicles were heavy and cumbersome. As automobiles increased in popularity it was necessary to speed up production and metal was introduced as a material which could be formed from large sheets by using dies.

The first panels were made with heavy tools in crude dies of sandbags and finished by hand at the bench. With the development of deep drawing steels, and with the demand to justify the cost of the expensive dies and machinery, large streamlined panels were designed and formed into contours that were more attractive, gave longer life and greater safety, and at the same time reduced the bulky construction previously required to give similar strength.

It is known that the shape of any material is held by the stresses set up in the material itself, in particular by those given by angles, crowns, channels, flanges, etc. The original shape will be maintained until the material is subjected to a force sufficient to overcome the initial stresses; furthermore, it will tend to return to its original shape providing that it has not been distorted beyond the point of elasticity. To illustrate this, take a sheet of light-gauge steel in both hands and apply force to the ends (Figure 1). The sheet

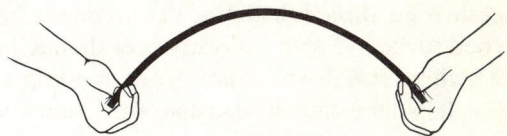

Figure 1 *Flexing of sheet metal*

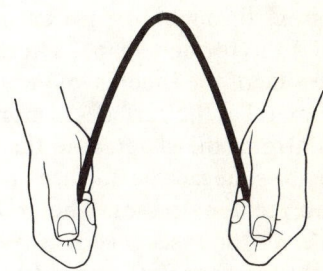

Figure 2 *Sheet metal bent beyond the limits of elasticity*

will flex and on removal of the applied force the sheet will return to its original shape. If, however, the sheet is bent beyond the limits of elasticity (Figure 2) it will take on a permanent 'set' and will not spring back. Stresses will have been set up at the bend greater than the original, and the material will have been 'work-hardened' at the bend.

It will be found also that when the applied force is reversed to straighten the sheet metal, it will not return to its original shape. It will now be bent in three places, there being a curve or bend at each side of the peak owing to the fact that having been work-hardened the steel at the bend is harder or stiffer than the adjacent areas which were not bent (Figure 3). If, however, instead of reversing the bending force, force was applied at the peak to release the stresses set up at the point of work-hardening, the sheet would spring back to its original flat state with very little distortion.

This building up of stresses at a bend or peak is also an important consideration in the design and manufacture of motor car bodies, the most common features of which are curved surfaces. These are called 'crowns' and may be curved either in one direction or in all directions. A crowned surface (Figure 4) is stronger than a flat sheet; whilst it will resist any force tending to change its shape and it also has the ability to return to its original shape unless distorted beyond its point of elasticity. These features are very true when the sheet has been formed in a press to give a permanent shape, with die-formed stresses throughout its entire area tending to hold the formed shape. On the bend or crown, one side of the sheet is longer than the other and the metal at the surfaces is more dense than at the centre of the sheet. The final action of the press is to squeeze the surfaces together, thus setting up greater stresses. The greater the crown, the greater the resistance to any change. A 'low crown' (a surface with very little curve), such as with some door panels, bonnets or flat parts of roofs, are springy, whilst 'high crowns' (surfaces with a lot of curve), such as wings or edges of roofs, are very resistant to an outside force.

A further method of giving strength to metal is to form angles or flanges along the edges of sheets. A right-angle bend greatly increases the strength of a sheet. Use of this method is to be found on inner door panels and also at the edges of wings, bonnets and boot lids where stiffness is required to unsupported edges.

Another method of providing strength is the use of U channels which, as the name implies, consist of two right-angle bends, one at each of opposite sides of a piece of metal; these channels are mainly used in the inner construction. Yet another method of increasing strength is known

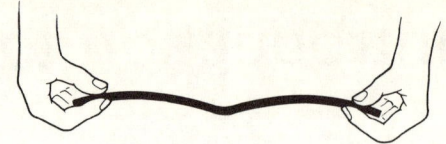

Figure 3 *Work-hardening*

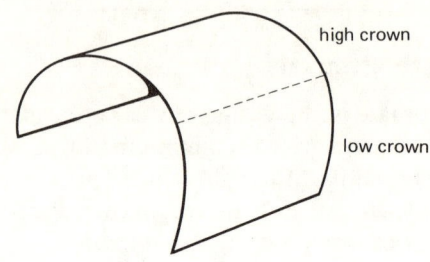

Figure 4 *Crown surface*

as 'box construction', which consists of two U sections welded together to form a square pillar or box. All these methods can be readily seen by inspection of any motor car body, and a little time spent in studying body design to ascertain where and how strength has been achieved will prove most profitable (see Figure 5 and Figure 6).

Insurance

All work of any nature should be done well, otherwise it is not worth doing, but when it comes to repairing an accident damaged car in order that it can go on the road again then the work becomes a matter of life and death. A badly repaired car is just another accident waiting to happen.

With more and more cars, faster cars, and not enough space for them, and with poor weather conditions, inevitably there are collisions. At the same time, costs are rising and the motoring public and insurers are becoming increasingly concerned about the repair bills they have to face. There is keen competition in the motor vehicle body repair trade and this is certain to

Introduction to body repair 11

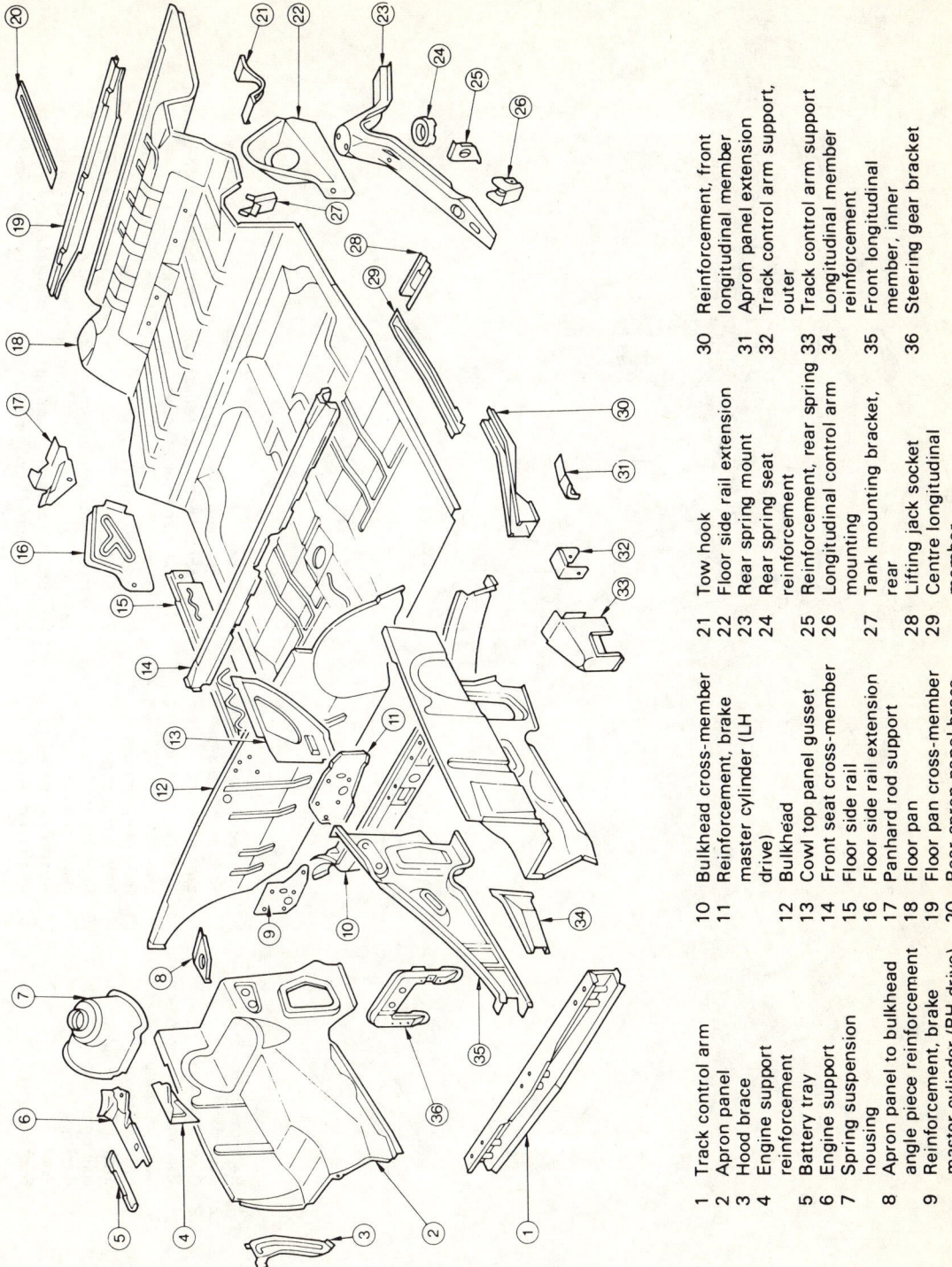

1 Track control arm
2 Apron panel
3 Hood brace
4 Engine support reinforcement
5 Battery tray
6 Engine support
7 Spring suspension housing
8 Apron panel to bulkhead angle piece reinforcement
9 Reinforcement, brake master cylinder (RH drive)
10 Bulkhead cross-member
11 Reinforcement, brake master cylinder (LH drive)
12 Bulkhead
13 Cowl top panel gusset
14 Front seat cross-member
15 Floor side rail
16 Floor side rail extension
17 Panhard rod support
18 Floor pan
19 Floor pan cross-member
20 Rear apron panel brace
21 Tow hook
22 Floor side rail extension
23 Rear spring mount
24 Rear spring seat reinforcement
25 Reinforcement, rear spring
26 Longitudinal control arm mounting
27 Tank mounting bracket, rear
28 Lifting jack socket
29 Centre longitudinal member
30 Reinforcement, front longitudinal member
31 Apron panel extension
32 Track control arm support, outer
33 Track control arm support
34 Longitudinal member reinforcement
35 Front longitudinal member, inner
36 Steering gear bracket

Figure 5 *Underbody assembly*

12 *Principles and Practice of Vehicle Body Repair*

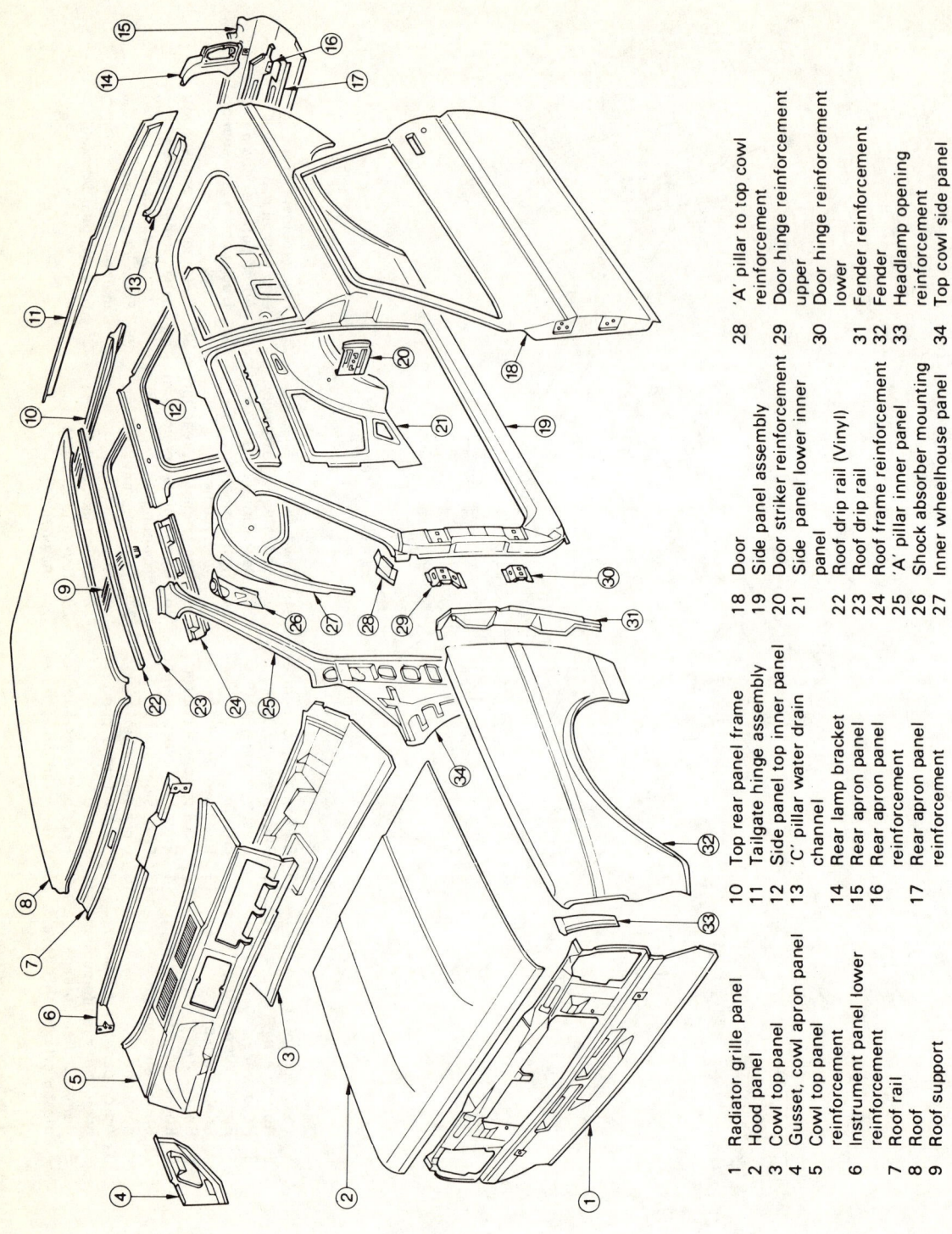

1 Radiator grille panel
2 Hood panel
3 Cowl top panel
4 Gusset, cowl apron panel
5 Cowl top panel reinforcement
6 Instrument panel lower reinforcement
7 Roof rail
8 Roof
9 Roof support
10 Top rear panel frame
11 Tailgate hinge assembly
12 Side panel top inner panel
13 'C' pillar water drain channel
14 Rear lamp bracket
15 Rear apron panel
16 Rear apron panel reinforcement
17 Rear apron panel reinforcement
18 Door
19 Side panel assembly
20 Door striker reinforcement
21 Side panel lower inner panel
22 Roof drip rail (Vinyl)
23 Roof drip rail
24 Roof frame reinforcement
25 'A' pillar inner panel
26 Shock absorber mounting
27 Inner wheelhouse panel
28 'A' pillar to top cowl reinforcement
29 Door hinge reinforcement upper
30 Door hinge reinforcement lower
31 Fender reinforcement
32 Fender
33 Headlamp opening reinforcement
34 Top cowl side panel

Figure 6 *Body assembly*

increase still more as the motorist (and the insurance company) is on the lookout for the lowest estimate. Repairing the damaged vehicles forms a vital part of the profits for a large number of garages but this work must be done efficiently to obtain maximum profits. It is the motor body that takes the brunt of the damage in many accidents and an impressive range of advanced motor body repair equipment is available to help the repair garage with this work. In the face of ever tougher requirements, both technical and economical, the garage has to employ the most efficient methods and the latest tools and equipment if it is to be in a strong enough competitive position to flourish.

In collision damage repairs there are frequently two conflicting interests. Usually the bill is met by the insurance company (all vehicles must have a minimum of third-party cover whilst a large amount of vehicles have fully comprehensive insurance cover), so that on the one hand we have the owner who wants to get as much done as possible for the premiums he has paid, and on the other hand we have the insurance assessor who wants to secure the most economical repair possible consistent with a satisfactory job. Maintaining the balance between these two we have the body repairer, who must use all the skill and techniques at his disposal to satisfy the customer and at the same time keep the repair bill as low as possible.

Throughout industry today – and the motor body repair trade is a major industry – the accent is on greater productivity through increased efficiency, and this can only be achieved by modernization. Greater productivity brings lower unit cost – the same job at a cheaper price – and it is the competitive estimate that gets the order every time.

When a vehicle is taken to the body repair shop the first thing the workshop manager needs to do is to determine who is going to pay for the repairs; if a customer is to pay out of his own pocket, then an estimate is prepared for approval or otherwise. When a vehicle is involved in an accident resulting in collision damage, the insured customer will make out an accident claims' form from his insurance company. This form deals with such factors as the nature of the accident, the details, and where the vehicle has been taken for body repair. Many insurance companies authorize the work to start immediately but body repairers usually prefer to have their work programme authorized in writing before they start any work on the damaged vehicle.

The Motor Insurance Repair Research Centre

The Motor Insurance Repair Research Centre was founded in the early 1970s by leading insurance companies in the UK. At a time of very high repair costs, research carried out at the Centre has concentrated on achieving more economic repairs through increased efficiency. Other research is carried out in design and paintwork techniques. Standard repair times have been established for the replacement of most car body panels and a manual for this and one for paintwork have been produced. Insurance assessors use these references as guides for body repair times.

Proposals for tighter controls over the repair of accident damaged vehicles have been submitted to the Department of Transport, based on the premise that many damaged cars are inadequately repaired and are being returned to the road in a potentially dangerous condition. The proposals call for a system of two-tier registration, whereby only those repairers with approved standards of equipment and staff will be allowed to repair structurally damaged vehicles, the remainder being confined to minor accident damage.

Repair or replacement

In general the rectification of crash damage falls into three main categories:

1. The straightening, realigning and reshaping of damaged sections that are capable of being repaired economically.

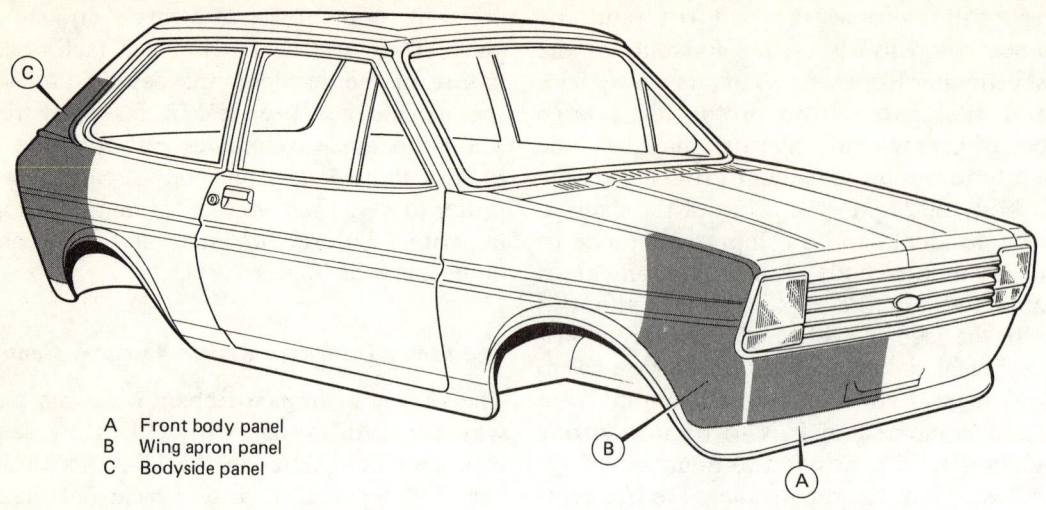

A   Front body panel
B   Wing apron panel
C   Bodyside panel

Figure 7 *Part-panel replacement*

2 The cutting out and replacing of severely damaged panels and members that are completely beyond repair (at least beyond economical repair).
3 Using part panels to replace damaged sections of body panels.

The panel or part-panel replacement method of repair (Figure 7) is often the most economical, owing to the availability and low cost of replacement parts, compared with the high labour costs of panel beating and straightening seriously damaged parts.

The decision on which of the three methods to use must first of all depend on the type and extent of the damage, i.e. whether it is physically possible to repair a particular section, though even this decision will be affected by the equipment available. Having decided that repair is possible, the next question is whether that repair is going to be cheaper than replacement.

The first essential step in the rectification of accident damaged car bodies is to make a thorough and complete assessment of the damage involved. If the damage is other than minor in nature, then in addition to direct visible damage caused by the impact, consequential or indirect damage can well exist in areas some considerable distance from the point of impact owing to hidden distortion to structural components.

It is a fundamental rule of body accident repair work that the repair operations must be carried out in the reverse order of the forces and stresses causing the damage. The assessment therefore should determine the cause and extent and also the sequence in which the damage occurred. If, during this assessment, damage or distortion which will affect body alignment is apparent or suspected, then it will be necessary to carry out a body alignment check, using purpose-built jigs. In the event of damage to front or rear end of the vehicle involving suspension or steering components, it is essential that the replacement parts are replaced in correct alignment to the undamaged body shell. For the alignment to be correctly carried out, it is essential that a body alignment jig is used to set the panels prior to welding.

It is generally accepted that the four corners of any vehicle are most vulnerable in minor collision incidents. The damage caused is invariably contained within the front headlamp, wing, front body panel and front wing apron, or within the rear quarter back panel areas, with the greater portion of each of the panels involved remaining in an undamaged condition. In many instances, present repair practice involves the purchase of complete body panels, leaving the operator the option of either replacing the complete panel or cutting a suitable section off the panel and discarding the remainder.

With the older type of car construction, employing a separate chassis and body, it was a relatively simple matter to unbolt or cut out damaged body panels and replace them with new ones. The panels themselves were generally fairly simple and therefore relatively inexpensive and the labour cost for replacing them was not high. Therefore, if the damage went beyond simple dressing of an easily accessible section, it was probably much cheaper and satisfactory to effect a straightforward replacement.

Chassis damage could be dealt with by the relatively unsophisticated body jacking equipment then available, or by cutting out badly damaged sections and welding back new pieces with gas cutting and welding equipment. The decision whether to repair or replace was quite simple as the dividing line between the two was fairly clearly defined and depended more on what it was possible to do than on what it was economical to do. Now, car construction has reached a much greater degree of complexity, and although practicability plays a large part in deciding the method of repair, relative cost plays a much greater part than before. Body panels themselves are more complicated in shape and therefore cost more and the cost varies even between makes of car depending on how the car manufacturer tailors the price of replacement parts. For this reason, a particular repair might be done more cheaply on one make of car by replacing a panel and on another make by repairing the damage *in situ*.

With monocoque or unit construction, which dispenses with a separate chassis, many body sections incorporate reinforcing members, and this not only increases their unit cost but also makes more difficult their removal and replacement. Exact positioning of the new section becomes highly critical, on occasions calling for expensive jigging to ensure the correct positioning of sections which will have to carry major mechanical components and suspension units.

Labour costs form a very high proportion of the total repair bill and much time and money is expended on stripping out to enable a section to be removed and on building up after replacement; this time and money is spent only to obtain access, not on the actual repair itself. It might be more economical to repair a damaged section than to replace it because a much higher percentage of the labour time is directed towards the actual repair activity, instead of to ancillary work, and, of course, the cost of the replacement part itself is eliminated. In any garage, the dividing line between repairing and replacing is decided not necessarily by what is the more economical but often by the scope of the equipment that is available. The customer will not always get the benefit of the most economical repair method because the shop may not be properly equipped. The customer is sometimes paying for the repairers' lack of efficiency, and in the end the customer must go to the better equipped workshop. This is inevitable as the insurance assessor, who is familiar with all the facilities available in an area, is bound to direct the work to the most efficient repair shop. With this in mind, the progressive equipment manufacturer is continually seeking new methods and techniques to push back still further this dividing line between repair and replacement and the repair industry must keep pace with this development to give the customer the benefit of this research.

Estimating

The first stage of the repair work is to make out an estimate for the damage and then to contact the insurance company so that their assessor engineer can visit the workshop to examine the damaged vehicle. An estimate has to be competitive to be acceptable; also it will be subjected to the careful scrutiny of the insurance assessor. For these reasons, it has to be accurate from the start and should leave no margin for error. The whole financial success of the body shop will depend on the skill of the estimator, assuming that he is backed up by first-class craftsmen with up-to-date equipment at their disposal. In many smaller body shops there is no separate estimator, and the man who prepares the estimate will also carry out the repair work; in such cases there is no lack of co-ordination between the two functions. However, in the larger organizations, where estimating is carried out by a separate individual, the need for this co-ordination and understanding can be seen quite clearly and is of paramount importance. The estimator must have full understanding of the scope of the equipment employed in the shop as this will materially affect his verdict when it comes to deciding whether to repair or replace.

In estimating there is naturally a tendency to lean towards the latter as, on the face of it, it seems the simpler method of repair and the preparation of the estimate is more straightforward. Also, being more straightforward, it is less liable to errors and easier to substantiate to the assessor. However, it does not necessarily produce the most competitive estimate and its choice can depart from the first principle of good estimating – to choose the most economical method of repair consistent with the production of a first-class job.

It is of the utmost importance that the estimator is as well versed as (if not in advance of) the actual repairer in knowledge of modern techniques. This also applies to the equipment he uses himself in examining the car for structural damage which may not at first be apparent. Distortion to the chassis or reinforcing

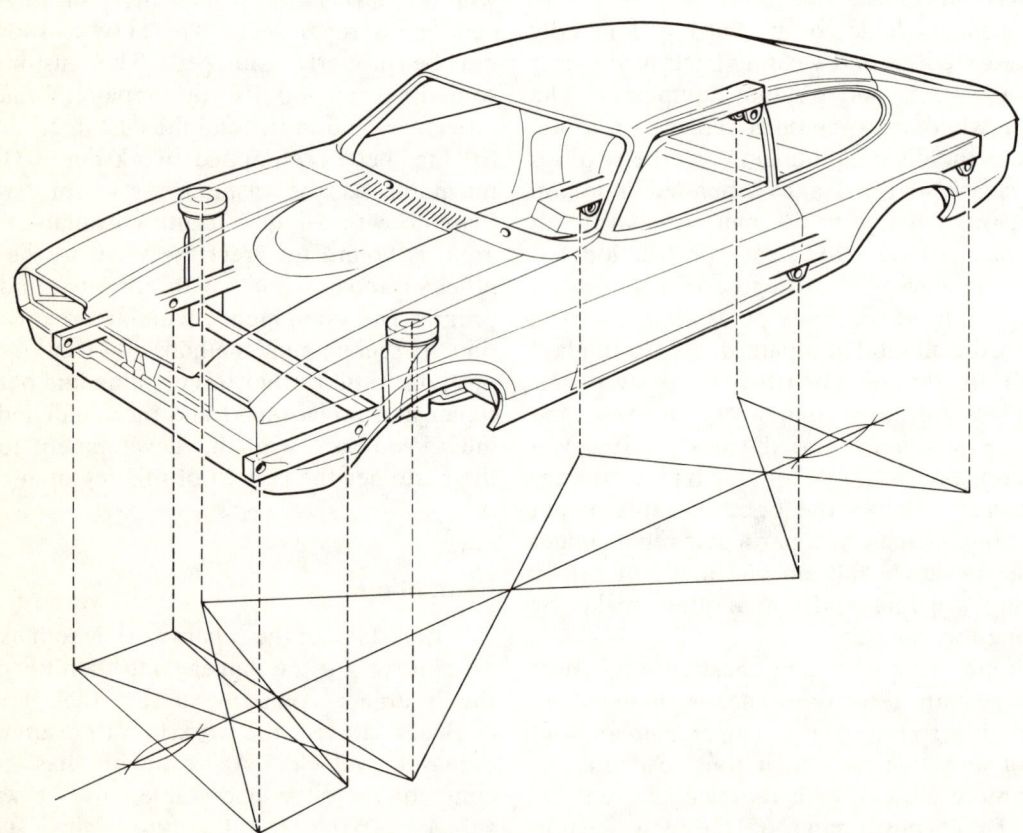

Figure 8 *The dropline method of alignment using five pairs of locating points*

sections of the body shell must be instantly recognized as this is the crux of the whole repair. In cases of bad damage this will be seen instantly, but it may be less obvious where there seems to be only superficial panel damage, unless a full drop test is carried out.

A drop test (Figure 8), in itself, is by no means a five-minute job if carried out properly, and is an expensive method of finding out that no structural damage is present. The use of proper frame gauges (Figure 9) will achieve the same results, with comparable accuracy, in a matter of minutes, showing an appreciable saving in the cost of the actual preparation of the estimate itself. Should a dispute arise with the assessor, it is only a matter of a few moments work to offer up the gauges to the car and prove the point. When the customer calls to collect the finished job and sign the certificate of acceptance, it is possible to build up considerable confidence by demonstrating with the same gauges the accuracy of the realigning work that has been carried out. The certificate of acceptance, sometimes referred to as a satisfaction or completion note, is a part of every accident insurance repair job. It is required by the insurance company to make it unlikely that a client will return several months later complaining that something was not done or that something else has arisen as a result of the previous accident.

One point should not be overlooked. In any factory, the inspector keeps his own gauges and measuring instruments, and these are never allowed out on the shop floor. If similar gauges are needed to check work in progress, a separate set is used, and the two sets are never allowed to be mixed. This is sound engineering practice and should be employed in all body repair shops; the estimator should keep his own master set of frame gauges which should never be used for shop floor work, except to check the accuracy of the shop gauges.

The work carried out by the estimator is all important; everything stems from his decisions and the whole pattern of the work of the shop has its origin with him. The work of the best repairer in the world can be negated by poor estimating, and inadequate understanding of the scope of the equipment and techniques available can lead only to uneconomical operation.

Requirements of body repair shop

A good body repair shop needs to have highly skilled craftsmen and good tools and equipment. The essential requirements can be listed as follows:

1. A well-planned layout operating on a flow line principle if possible, with repairs organized so that major work is completed in one area, minor work in another, sanding and flatting in another, then painting and finally building up and preparation for the customer (see Figure 10).

2. Alignment jigs for body and chassis checks (Figure 11).

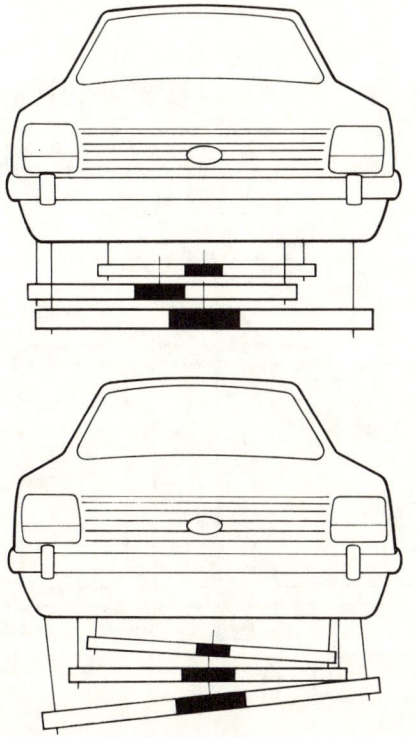

Figure 9 *Gunsights show how car underbody is misaligned*

18 *Principles and Practice of Vehicle Body Repair*

Figure 10 *Typical layout for body repair and refinishing shop*

3 Portable powered tools such as grinders, sanders, shears, saws, etc.
4 Pullers and dozers to pull out or push in damaged sections.
5 Sheet metal cutting and forming tools such as guillotines, folders, light presses and air chisels.
6 Comprehensive panel beating tools such as dollies, spoons, hammers, etc.
7 Welding equipment to include oxy-acetylene, metallic arc, resistance welding machines and/or metal arc gas-shielded welding machines.
8 A well-planned spray booth adjacent to a suitable low-bake oven.
9 Fire-fighting equipment.

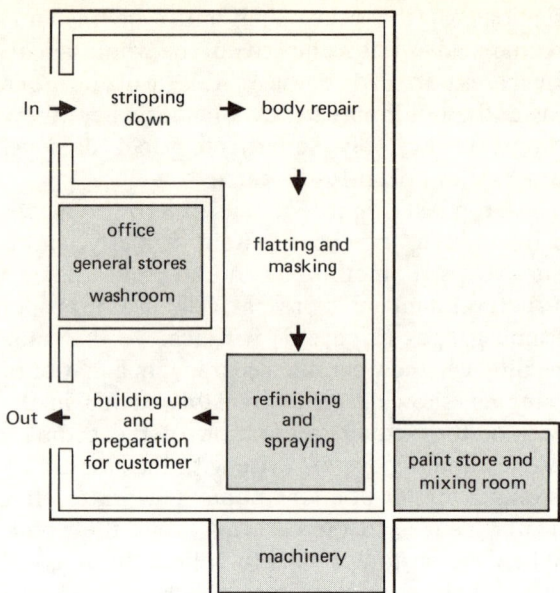

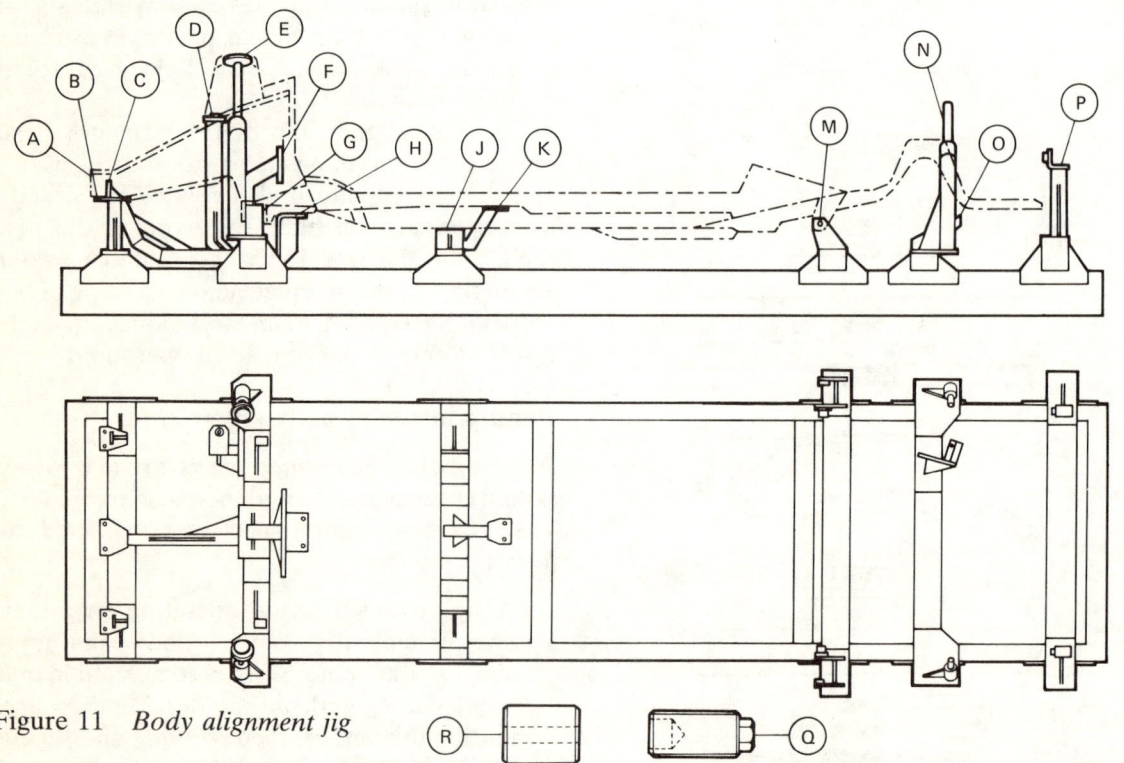

Figure 11 *Body alignment jig*

A Cross-member stiffener mounting
B Tie box mounting
C Engine support mounting (front)
D Engine mounting
E McPherson strut mounting
F Steering rack mounting
G Lower radius arm mounting
H Engine support mounting (rear)
J Side member sockets
K Selector housing
M Lower arm mounting
N Shock absorber mounting
O Panhard rod mounting
P Rear bumper mounting
Q Sleeve
R Sleeve

Introduction to body repair 19

Figure 12 serves as an indication of the body repair tools needed to make professional repairs; the range may be extended according to workshop requirements.

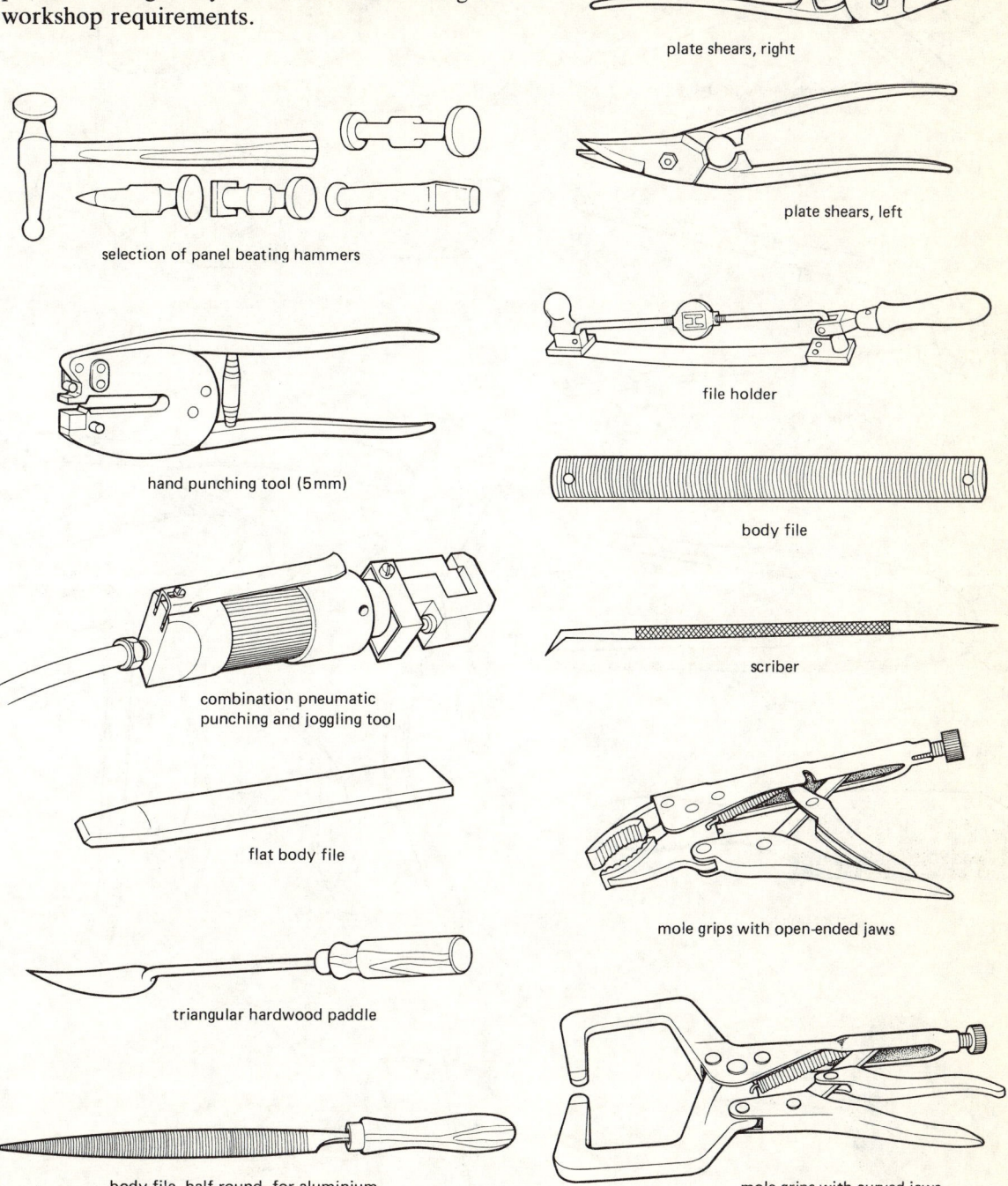

Figure 12 *Body repair tools*

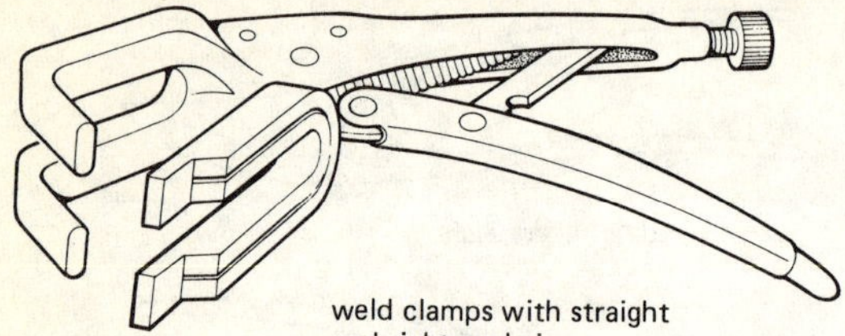

20 *Principles and Practice of Vehicle Body Repair*

weld clamps with straight and right-angle jaws

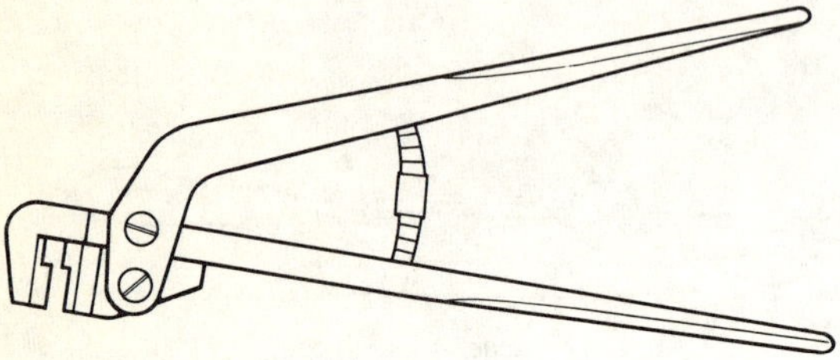

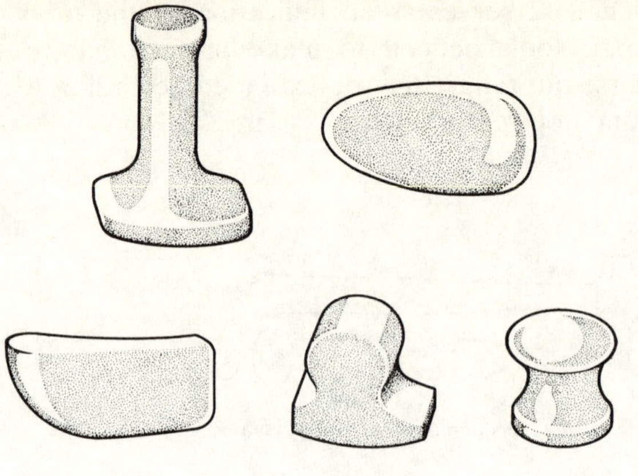

dollies of various shapes

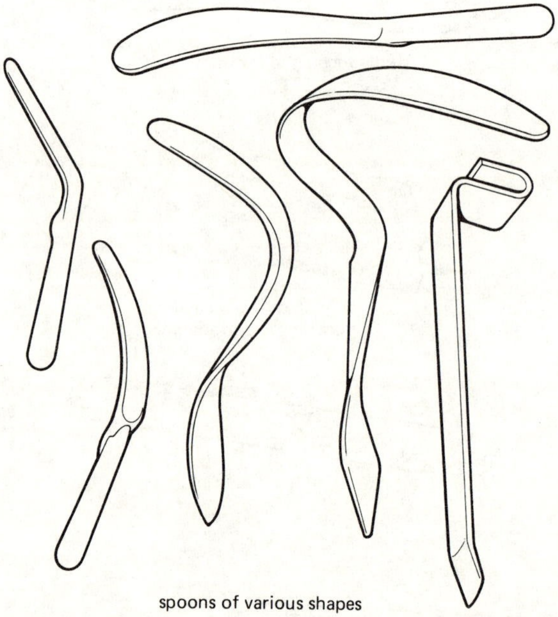

crimping tool

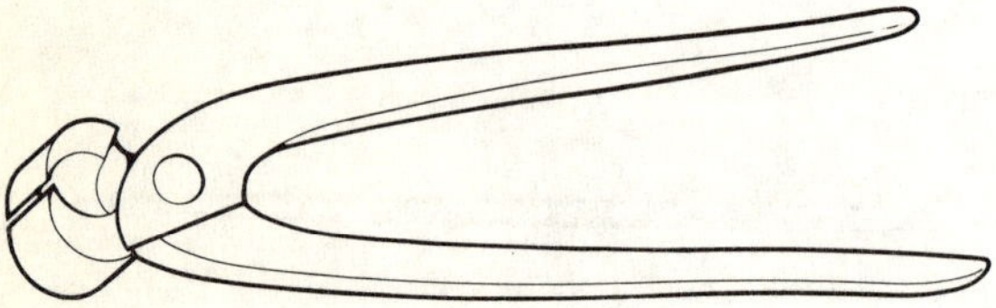

pincers

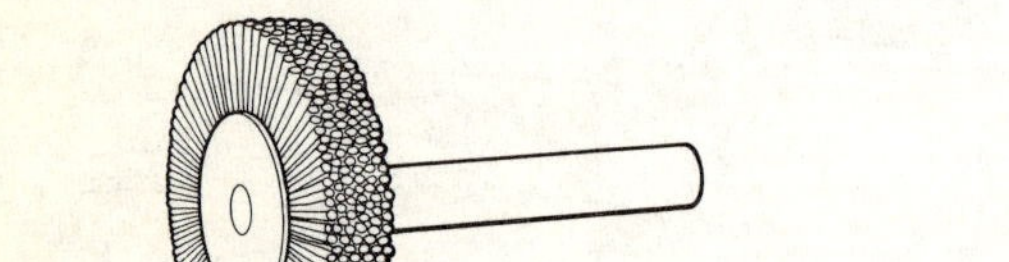

spoons of various shapes

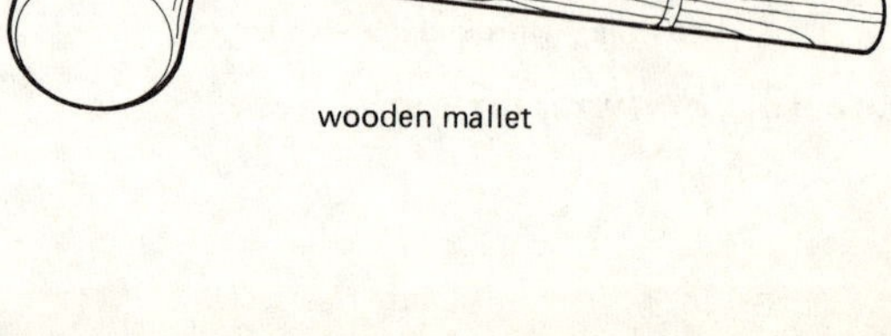

wire brush

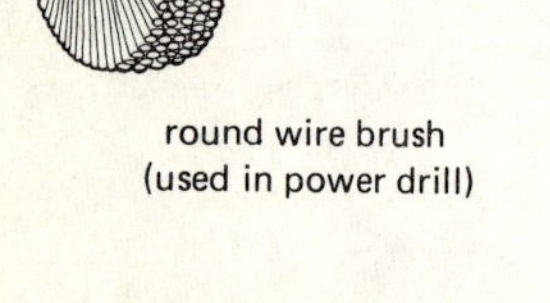

round wire brush (used in power drill)

wooden mallet

Introduction to body repair 21

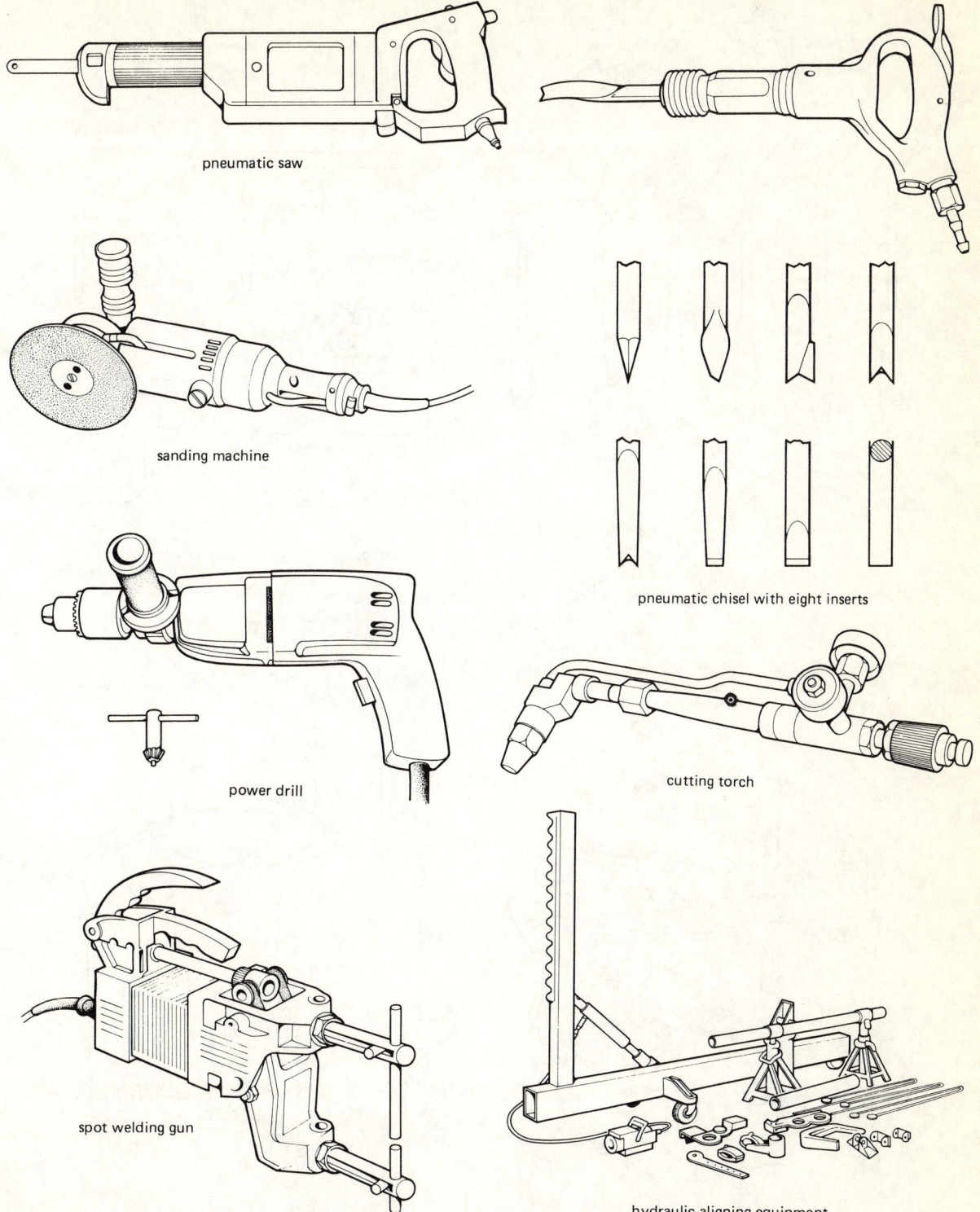

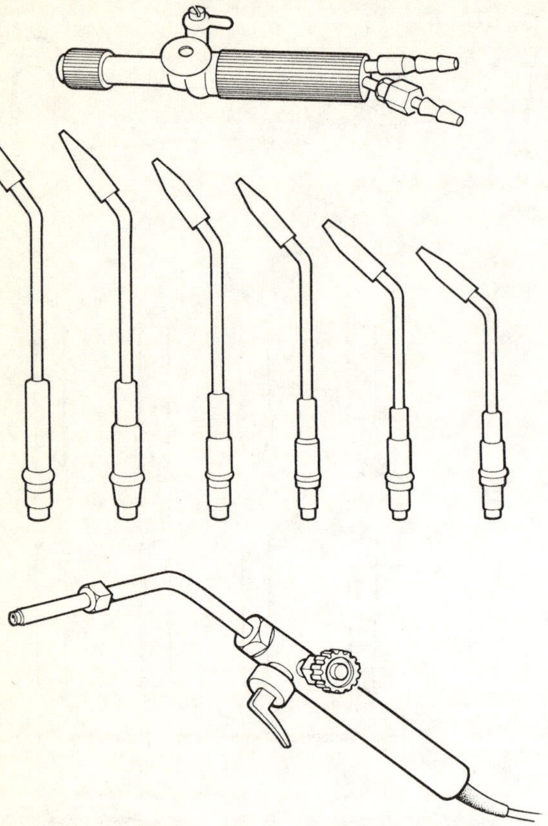

22 Principles and Practice of Vehicle Body Repair

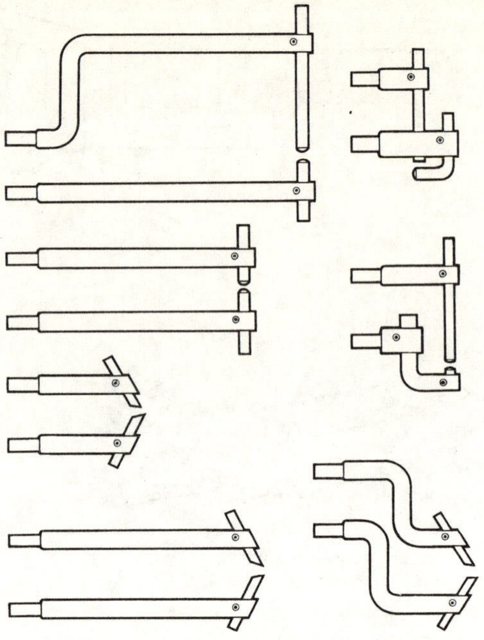

electrode arms of various shapes with electrodes

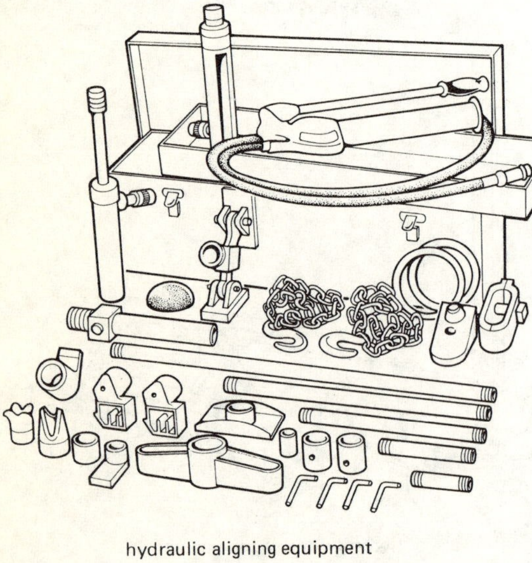

oxy-acetylene welding equipment with six burner inserts

hydraulic aligning equipment

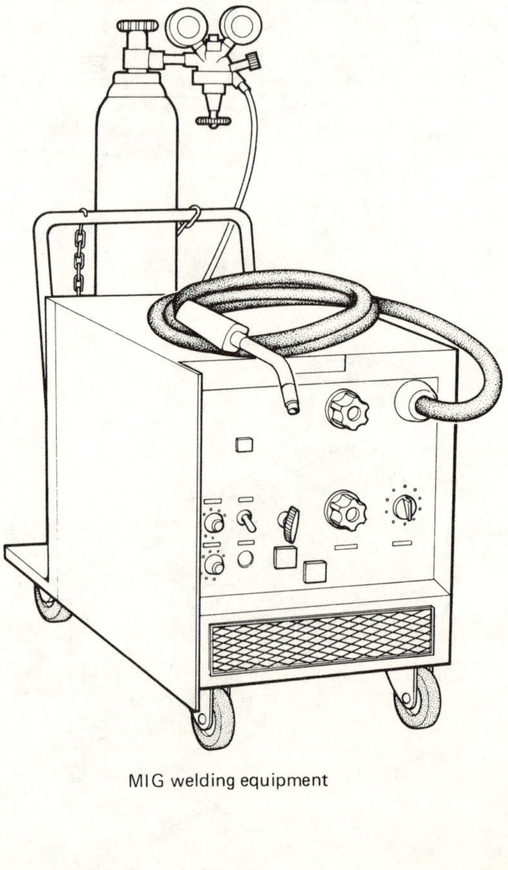

MIG welding equipment

Introduction to body repair 23

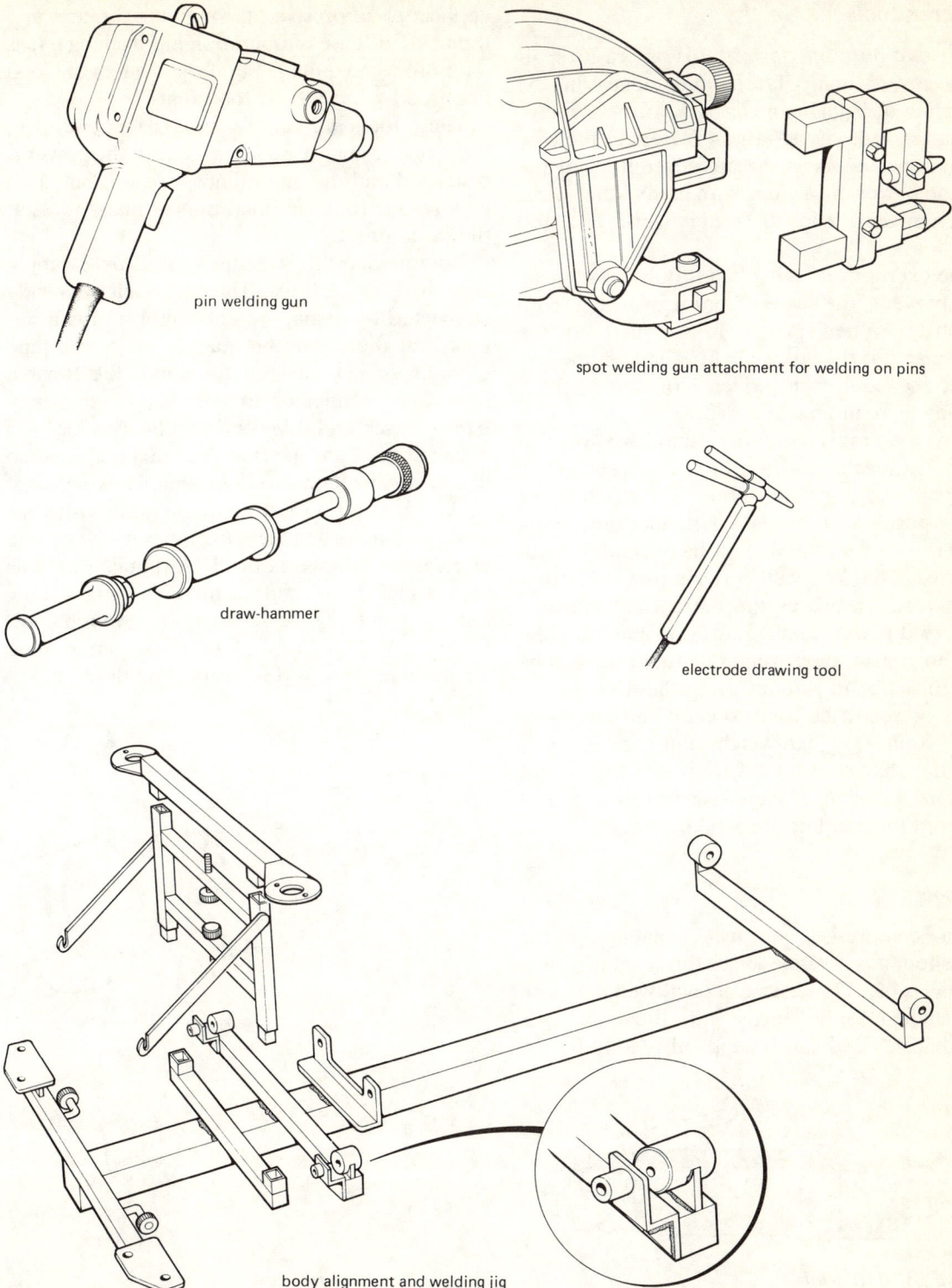

pin welding gun

spot welding gun attachment for welding on pins

draw-hammer

electrode drawing tool

body alignment and welding jig

Powered tools

The use of portable power tools on crash repair work covers many facets and stages in the operation of bringing a car back to the 'as new' condition. One of the largest factors in the case of car repair work is the labour content of the job and power tools along with many other items of equipment lead to better work and job handling.

The exertions of using hand tools cause tiring and prevent the worker from working continuously. When power is supplied to the craftsman it is then possible for him to guide and direct the equipment which works for longer periods than an unaided man.

Powered tools are the simplest way of providing mechanical aid to the worker. They are small and portable so that they can be used in any position. There is a particular advantage in their use when the work is heavy and difficult to move. The disadvantages are that such tools are not so precise as the more rigid machine tools, and that available power is limited by the need to restrict their weight so that they can be used in the hand without tiring the user.

For power to be used to exert the necessary force from this lightweight equipment, it is essential that operating speeds are high. This suggests that danger can arise from the use of powered tools unless users take care.

Air tools

When compressed air is available in the workshop, it is possible to use this form of power in small tools. These are compact, well balanced and shaped to fit snugly into the hand (see Figure 13). Feed can be applied readily by the application of pressure through the forearm and thumb. This ease of operation has its dangers, as air motors can quickly be stopped when the load imposed by the feed is too great.

These tools are very sensitive to air pressure; if the operating pressure is reduced, the power is restricted and the operating speed will fall. This may result from the operation of many tools at the same time.

Compressed air is easily wasted by escaping from bad connections. The tool itself is quickly connected by a snap-on connector and will not leak, but there may be many joins in the pipe carrying the air around the shop; the flexible hose from connector to tool has to be firmly fixed at each end if leaks are to be avoided. Air does not fall onto the floor and make a mess, so it can escape unnoticed in a noisy workshop. Make sure you do not waste air when using air tools, and report any leaking pipes which no one else seems to have noticed. Although air tools have a high power output for their weight, they are the most expensive form of powered tool. Only about 20 per cent of the power put into the air when compressed is exerted by the tool.

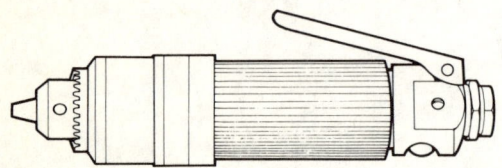

Figure 13   *Air tool*

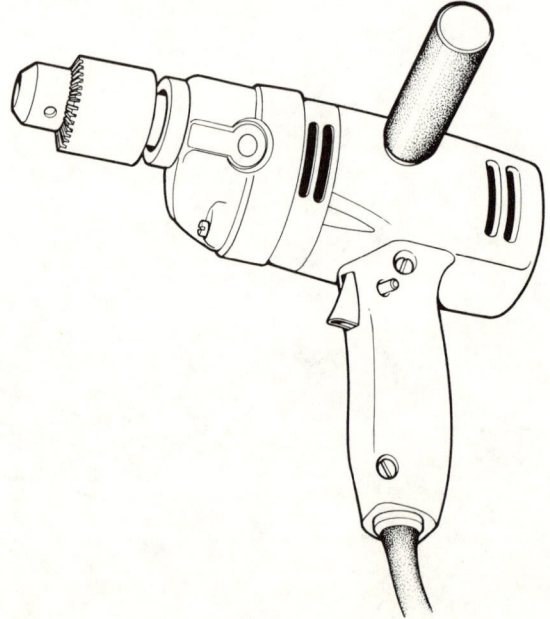

Figure 14   *Electric tool*

Electric tools

Compared with air tools, electrically powered tools (Figure 14) are much heavier and bulkier but they are more efficient – about 60 per cent as against 20 per cent. The supply voltage is constant so there is no falling off of power with fluctuations of supply, but the operating speed of electric tools drops rapidly under load. Working speed is usually about 60 per cent that of idling speed but will be reduced further if excessive feed is applied. With misuse the motor will stop and may burn out.

The ease of plugging in electrical tools may hide the fact that there is danger in operating such equipment. This is particularly acute when a high-voltage source of power is used. Electric tools should preferably be used on a 110 V circuit rather than on 240 V even if this means having a portable transformer when the supply is not at that voltage. Electric tools must never be connected to a 440 V circuit.

A further danger arises when tools are operated in wet conditions. It is quite possible for a killing current to flow if a person is part of a circuit of as low as 50 V under wet conditions.

Electricity is the cheapest form of motive power for small tools and because it is so widely used it has been responsible for many accidents involving shock from a metal casing of the tool which has become live. Most power tools now available are manufactured using the double-insulated principle. This is a system whereby the motor and other live parts are isolated from any section of the tool that it is possible for the operator to touch. Double-insulated tools bear the BS 2769 kitemark and the double-squares symbol as shown in Figure 15 and do not require an earth wire.

The use of extension cables with socket outlets should be avoided when possible. There is no protection against serious shock from a damaged cable; the only safeguard is to make sure that the flexible cables carrying the power supply are undamaged before tools are used.

Both types of powered tool can be used for a variety of industrial operations. Of these the most common is the production of a hole by drilling. The electric 13 mm variable speed drill has a two-speed mechanical gearbox which provides extra power at low speeds. The variable speed facility allows the operator to select the correct drilling speed for various materials. This is particularly valuable in the body repair shop where requirements include drilling holes to accommodate aerials, mirrors, etc. and for drilling out spot welds (see Figure 16).

Power tools can also be used for cutting metal, particularly in thin-plate metal working. The cutting is performed by a to-and-fro movement of the blade. There are two types of action:

1 By cutting like a guillotine.
2 By a punching action, called nibbling.

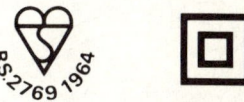

Figure 15 *Kitemark and double-squares symbol*

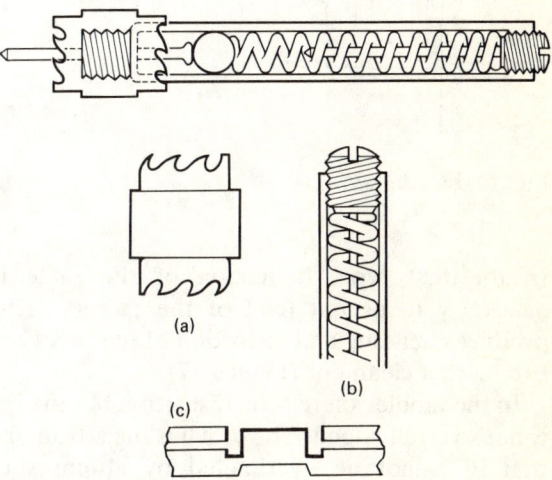

Figure 16 *Zipcut spot-weld remover, used with electric or air drill. Cutter blade (a) is reversible. Adjustment (b) provided for varying depth of cut so that only upper panel is released (c)*

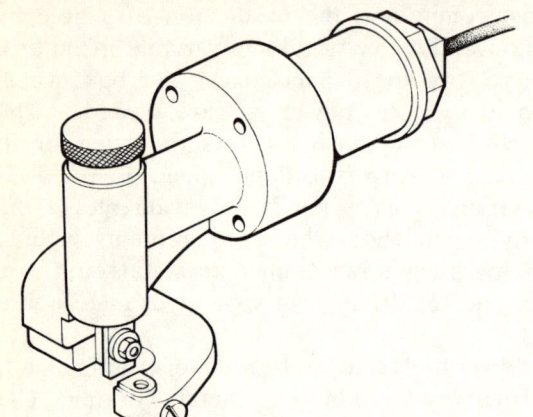

Figure 17 *Power shear*

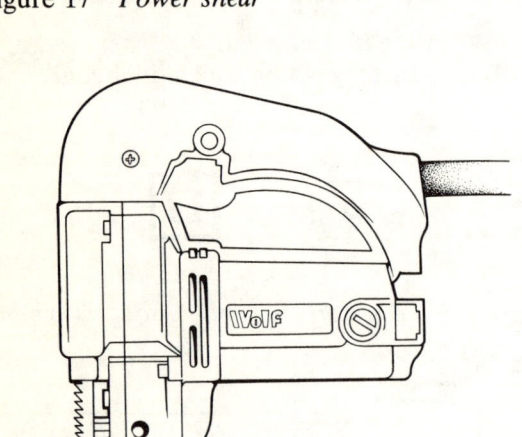

Figure 18 *Jig saw*

Figure 19 *Straight grinder*

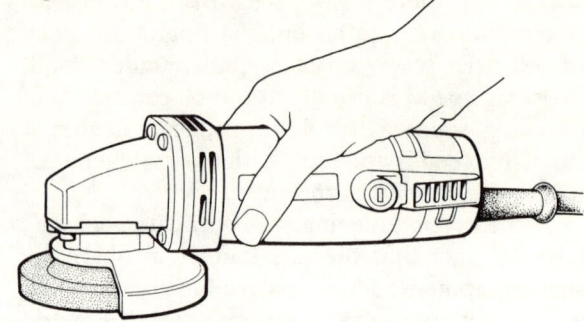

Figure 20 *Grinderette*

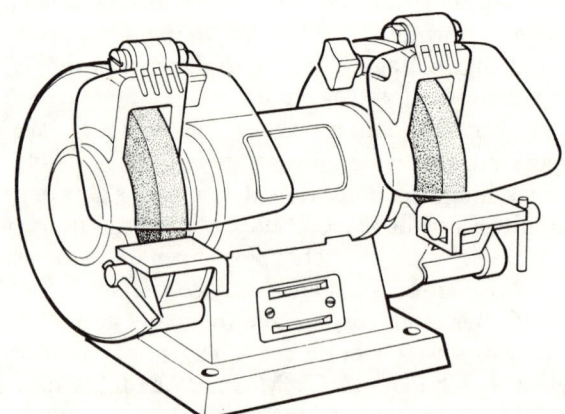

Figure 21 *Bench grinder*

In the first type, deflection of the plate is necessary to permit feed of the cutters. This produces bending and distortion of the sheet but produces a clean cut (Figure 17).

In the nibbler there is no distortion but the cut is not so clean edged. The tool has the advantage that it cannot be overloaded by attempts to overfeed it. In such a situation the power shear will stall and refuse to cut. These cutting tools can be used for speedy and accurate cutting of unusual shapes without regard to the area of metal worked on, since they are portable and can be made to work at any point of even the largest sheet. A portable jig saw (Figure 18) is a very useful item in the body repair shop for cutting hard and soft wood, low-carbon steel, aluminium, acrylics, stainless steel, etc. There are a range of blades available suitable for the various materials.

The 120 mm straight grinder (Figure 19) is ideal for grinding or wire brushing in tight spaces such as circular cut-outs in sheet metal, etc.

The grinderette (Figure 20) is an extremely versatile one-hand grinder. Used with appropriate discs it can grind, cut, trim and sand a wide range of materials. In the repair shop the grinderette can cut away damaged body panels, remove rust, dress spot welds, and prepare surfaces for filling, priming and respraying.

A bench grinder and polisher (Figure 21) are a must in any repair shop for sharpening small tools, grinding small components to make adjustments, shaping, etc.

The portable angle disc sander (Figure 22) is ideal for the sanding of large areas and the heavy duty model can, with the appropriate conversion pack, be converted into an angle grinder.

One of the most regular jobs in any body repair shop is shaping and smoothing filler. The finishing sander (Figure 23) ensures a fine finish with 12,000 full 3 mm diameter orbits per minute. This prevents the build-up of harmful particles on the sanding sheet which could otherwise scour the work surface.

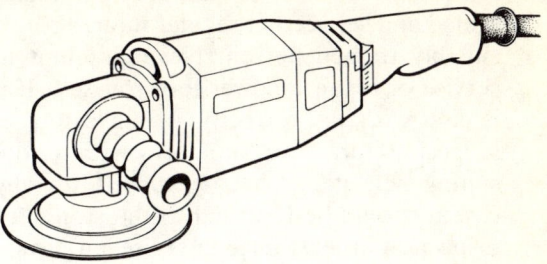

Figure 22   *Disc sander*

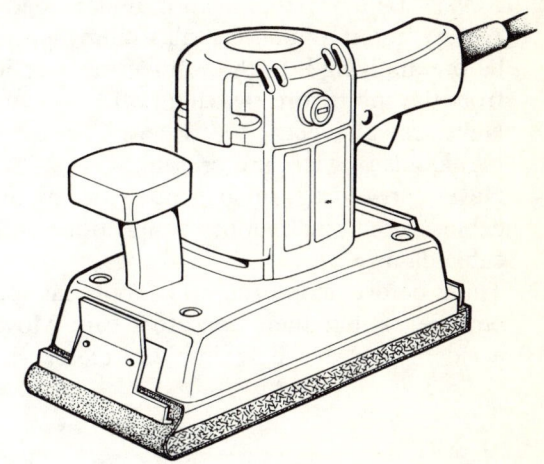

Figure 23   *Finishing sander*

Safety

Although each type of power tool has its own individual safe working procedures the following basic safety rules should be followed when using any powered tool:

1. Never use a power tool unless you have been properly trained in its use.
2. Never use a power tool unless you have your supervisor's permission.
3. Always select the correct tool for the job you are doing (if in doubt consult the manufacturer's handbook).
4. Ensure that the power supply is correct for the tool.
5. Ensure that the tool's cable is:
    (a) free from knots and damage;
    (b) firmly secured by the cord grips at both the plug and tool ends;
    (c) unable to come into contact with the cutting edge or become fouled during the tool's operation. This can be done by draping the cable over the shoulder during operation.
6. Before making any adjustments always remove the plug from the socket. Also ensure that the tool is switched off before replacing the plug in the socket.
7. Always use the tool's safety guards correctly and never remove or tie them back.
8. Never put a tool down until all rotating parts have stopped moving.
9. Always wear the correct protective equipment for the job. These may include:
    (a) safety goggles;
    (b) dust masks;
    (c) ear protectors;
    (d) safety helmet.

Note: Loose clothing and long hair should be tied up so that they cannot be caught up in the tool.

10 All power tools should be properly maintained and serviced at regular intervals by a suitably trained person. Never attempt to service or repair a power tool yourself. If it is not working correctly or its safety is suspect, return it to the stores with a note stating why it has been returned. In any case it should be returned to the stores for inspection at least once every seven days.
11 Ensure that the material or workpiece is firmly clamped or fixed in position so that it will not move during the tool's operation.
12 Never start or stop a tool under load. Always let it obtain its maximum speed before applying it to the job and remove it from the job before switching off.
13 Never use an electric tool where combustible liquids or gases are present.
14 Never carry, drag, or suspend a tool by its cable, as this causes loose connections and cable damage.
15 Think before and during use. Tools cannot be careless but their operators can. Most accidents are caused by simple carelessness.

Health and safety

Workshop safety and accident prevention is of paramount importance and must be given every consideration at all times. Carelessness and the disregard of safety precautions cause accidents which can have serious far-reaching effects on the victim and his family. Accidents, as well as being painful and unpleasant, result in much lost time and money and can cause serious injury, sometimes of a permanent nature.

Dress and behaviour

The most suitable wear is a one-piece overall or boiler suit in good condition. Torn material can catch in moving machinery. Overall buttons must be kept fastened and ties and scarves should not be worn, for – like torn material – they can catch in moving machinery. Where there is machinery with moving parts, sleeves should be tightly rolled up above the elbows, and where it is necessary to protect the skin, closely fitting sleeves should be worn down to the wrist.

Loose bandages should not be worn and it has been proved dangerous to wear a ring or a watch.

Safety boots or shoes must be worn. Goggles should be worn to protect the eyes. Failure to wear any of the safety clothing necessary for a particular job may result in serious injury.

Hair must be kept short or a cap must be worn.

Barrier cream should be applied before work to prevent chemical dermatitis.

Safety with tools

Sharp tools must always be carried with their points downwards and, where possible, with the sharp edges protected. Sharp tools and marking out and measuring instruments, such as dividers and scribers, must not be carried in overall pockets.

Defective tools must never be used.

When working overhead, tools must be attached to safety lines.

Lifting and carrying loads

Serious injury can result from lifting loads which are too heavy. Help should be obtained when lifting anything very heavy, and lifting equipment, inspected before use, should be used where possible. In lifting, leg and thigh muscles should do the work. A firm stance should be taken with feet slightly apart to give good balance. The knees should be bent, the back kept straight and the chin tucked in. The load should be firmly gripped and lifted by straightening the legs. Sharp edges and slippery surfaces should be avoided. It must always be possible to see over the load and round the sides of the load being carried.

Introduction to body repair

General behaviour

Everyone must move about the workshop with care – never run.

The gangways must be kept clear.

Never try out or use machines unless authority is given.

Walking under suspended loads should be avoided.

Practical jokes are a serious danger and they are absolutely forbidden.

Housekeeping

The bench must be kept neat and tidy and only the tools needed for the jobs should be out. Every tool should have its proper place, and the floor must be kept clear and clean. Separate bins should be used for waste rags and dirt and spilt oil should be wiped up at once.

Electricity

No defective electrical equipment may be used and any faulty equipment must be reported immediately.

Electrical equipment must never be touched with wet hands. Frayed wires are dangerous and must be reported, and electrical connections must be made by an authorized electrician.

Electrically driven hand tools must be properly earthed. If the earth wire is not connected to the earth pin, the outer casing can become live and give a fatal shock. If in doubt about earthing a supervisor should be consulted. Power tools should never be connected to a lamp socket, as these have no earth connection. Use only double-insulated electrical power tools.

Trailing wires can cause an accident, and danger notices mean what they say.

Compressed air

Compressed air is very powerful and its misuse can be very dangerous. Air lines must only be used for the purpose intended. Compressed air can drive particles of dust and swarf through the skin and can cause blisters; it must never be directed towards the skin or clothes.

Machines

Take care when walking about in the vicinity of machinery. Before starting a machine, the operator must know how to stop it. Most machines must be switched off when not in use and if, for any reason, a machine such as a grinder must be left running and unattended then a clearly visible notice should be displayed. The main danger points when using machines are the moving parts, such as gears, belts, cutting tools, saws, presses, grinders, etc. Never touch any moving part on a machine; swarf, hot metals and sharp edges on work are all dangerous. Swarf should be removed with a sash brush but only when the machine is stationary. All guards must be in place before a machine is started.

Repairs to glass-fibre reinforced plastics bodywork

The repair of any glass-fibre reinforced plastics moulding must be considered by basically applying the same technique as is used in producing original mouldings. When a moulding becomes damaged there is the economic position to consider. For example, if a panel is just slightly damaged by an object penetrating, an overnight repair is both possible and economical, but if the whole side panel is ripped right through owing to heavy impact, so that there is the possibility of part of it being rendered useless, there is little economic value in attempting a repair.

Repair of these mouldings may be accomplished with the aid of kits comprising resin, catalyst, accelerator and a selection of glass-fibre materials, such as random mat and woven cloth.

When the damage is extensive it is a sound method to back the repair with a large piece of resin-impregnated cloth, covering the extremes of the damage. First, the splits should be rubbed clear of the damaged resin and also to expose some of the torn glass fibres which will provide a key for the new material. The glass cloth should then be applied to the back of the damaged area and, after well impregnating with resin mix, left to harden; the time for this depends upon the actual mix employed. Next, a small quantity of glass fibre is well compounded with resin mix and this is then worked into the splits with a spatula or flat blade. If a hole has to be filled, the same system is used but the hole is filled with the glass/resin mix. If the hole is large it is advisable to fill with several layers, allowing each application to set up before applying the next.

Generally speaking, heat is not essential to set up the plastics material because the resin reacts, once the catalyst is added to it, so that the liquid solidifies. But it is advisable to carry out such repairs in a 'paint shop' atmosphere because damp or moist air will have its effect on the set-up. Excessive heat, on the other hand, will produce a brittle effect on the solid which will the more easily chip away from the glass fibres.

Where painting is required, all traces of mould release agent must first be removed and it is advisable to sand the part with medium-grit paper. Undercured mouldings and those with air voids near the surface can allow the paint to sink after a period. Experience is that synthetic paints or hot sprayed cellulose are preferable to air-drying cellulose.

Extreme care must be exercised and relevant safety precautions observed when handling GFRP materials.

2 Refinisher protection and safety

A healthy and safe environment is essential in any workplace and the value of the legal requirements, designed to preserve health and safety, will be appreciated by practical application with common sense as, probably, the greatest asset to working relationships. Development of healthy and safe conditions promotes operator satisfaction and efficiency and increases profitability; however, this is often overlooked, particularly when dealing with the hazards associated with refinishing equipment, materials and processes. Close co-operation and an understanding relationship with local council or government, manufacturers, and workforce is necessary. The following may be regarded as a framework on which a safety procedure can be based after discussion with local enforcement bodies and employee representatives.

Safety signs

Visual reminders on all aspects of safety serve as a constant reminder to all refinish operators. NO SMOKING, SHUT DOORS, WEAR FACE MASK, WASTE DISPOSAL POINT, FIRE ALARM, etc. are a few examples. In addition, bay areas and fire equipment points, clearly marked, promote a tidy working area, and safety instruction sessions and posters displaying regulations and basic rules for safety lead to a general awareness of *responsibility*.

The 1976 British Standards Institution (BSI) standard, BS 5378: 1976: *Specification for Safety Colours and Safety Signs*, divides safety signs into four categories each with certain characteristics of shape, colour and symbols used. Two further categories are included in the following list:

1. Warning signs (Figure 24) comprise a yellow triangle with a black outline. The symbol or text is also in black. This combination of black and yellow identifies CAUTION.
2. Prohibition signs (Figure 25) are recognized by a red circle with a cross bar running from top left to bottom right on a white background. The symbol is in black within the circle. The colour red is associated with STOP or DO NOT.
3. Mandatory signs (Figure 26) indicate that a specific course of action is to be taken. The colour is blue and the symbol or text is white.

Figure 24 *Warning signs*

Figure 25   *Prohibition signs*

Figure 26   *Mandatory signs*

Figure 27   *Safe condition signs*

Figure 28   *Supplementary signs*

Figure 29   *The new DANGER sign*

4 Safe condition signs (Figure 27) provide information for a particular facility and are rectangular with white text or symbol on green background. The safety colour green indicates ACCESS or PERMISSION.
5 Supplementary signs (Figure 28) containing text only are used in conjunction with a safety sign to provide additional information. The sign must be rectangular and coloured, with black text on a background of either white or a colour representing the 'safe condition'.
6 The new DANGER sign (Figure 29) introduced in BS 5378: 1980 is to be used to identify the perimeter of any hazardous area. It may be used with or without a safety sign and consists of yellow and black diagonal stripes.

Legal requirements for materials and employees

1 The Petroleum Consolidation Act 1928 and Statutory Rules and Orders 1929, no. 993 and 1947, no. 1443.
2 The Highly Flammable Liquids and Liquified Petroleum Gases Regulations 1972.

It is normal for paints governed by 1 to be labelled 'Petroleum mixture giving off an inflammable heavy vapour' (Figure 30).

Those governed by 2 must be labelled with the appropriate flash point, as either 'Flash point below 22 °C' or 'Flash point in the range 22–32 °C'. Paint users must understand the statutory requirements of the above Acts and the need to obtain a licence from the local authority for the storage of paint which is subject to the Petroleum (Consolidation) Act.

3 The Health and Safety at Work etc. Act 1974.

The safety requirements for vehicle refinishing are concerned with handling, storage and application of refinish materials, the correct siting and use of electrical equipment, and healthy working conditions in the shop, particularly with regard to spray booth design, air extraction and ventilation. The statute affects the manufacturers and suppliers of equipment and paint as well as refinish employers and operators.

The manufacturer and supplier has a duty to ensure that the safety of any refinish product or equipment is without risk to health when properly used. Warnings connected with the product composition, or special instructions for its use, may be printed on the container label or supplied in the form of product information literature. The Health and Safety at Work etc. Act 1974 imposes a duty on all persons at work.

Employers

Safe and healthy conditions of work are always the direct responsibility of the employers. This means that apart from the obligations laid upon the refinisher by the Act and by general factory legislation, the strict regulations governing the siting and operation of equipment must be observed and provision made for air extraction, ventilation and the safe storage of materials.

Employers must ensure the safety of their employees by maintaining safe plant, safe

Figure 30 *Flammable material signs*

34 *Principles and Practice of Vehicle Body Repair*

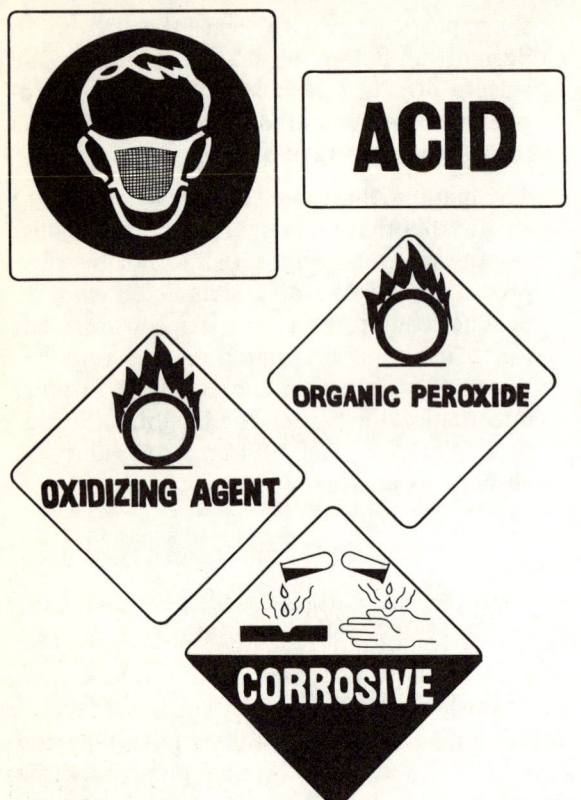

Figure 31 *Handling and use of materials and equipment*

systems of work and safe premises; they must also ensure adequate instruction, training and supervision.

Employees

It is the responsibility of the employee to practice common sense, to check up on safety precautions, read the manufacturer's data and note any warnings about the handling and use of materials or equipment (Figure 31). Employees have a duty to always wear protective clothing when required and to keep the work area clear and tidy. Carelessness or neglect can put colleagues at risk. All employees must take reasonable care in their work activities, to avoid injury to themselves and to others, and have an obligation to co-operate in meeting statutory requirements; this includes the responsibility to inform the management of any situation occurring within the refinish facility that may jeopardize safety or security of staff and/or premises.

General regulations for siting and workshop facility

1 The Control of Pollution Act 1974.
2 Deposit of Poisonous Waste Act 1972.
3 Local by-laws regarding pollution and public health.

If in any doubt the local factory inspector/environmental health officer should be invited to discuss needs relative to these particular subjects.

4 Fire regulations and basic requirements for the availability and maintenance of fire-fighting equipment.
5 Recommendations as contained in BS Code of Practice CP 1003, available from the British Standards Institute.
6 The Factories Act 1961 (Equipment).

Fire regulations and basic requirements for the availability and maintenance of fire-fighting equipment

Adequate fire prevention and fire-fighting equipment must be provided and maintained in all areas where paint is used, handled or stored. Consideration should be given to the installation of fixed automatic sprinkler systems to provide protection from fire in these areas. The location, uses and limitations of fire-fighting equipment must be familiar to all personnel. Extinguishers must be accessible and checked regularly. All staff must be familiar with relevant fire drill.

Precautions and control against sources of ignition

To minimize spillage, evaporation and hence fire risk, the lids of part-empty tins must always be replaced immediately. Various combinations of paint materials can be extremely hazardous;

these may occur in the wastebin. Always dampen waste materials with water; utilize separate covered containers for rag and waste paper to eliminate the possibility of spontaneous combustion. It is advantageous to compress waste paper into bales where an appropriate press is available. Remove all waste materials at the end of each day.

Certain catalysts in common use contain organic peroxides which are powerful oxidizing agents and are therefore potential causes of ignition (Figure 32). Containers of such materials *must* be filled with water when empty and prior to disposal.

Solid residue, i.e. overspray dust, resulting from the use of cellulose paint is also highly flammable. Reference must be made to the latest regulations governing the use and storage of these materials. Dipsticks, stirring instruments or scrapers must be of stiff fibre, non-ferrous metal or non-sparking material.

Static electricity: electrical discharges
In particular, the installation and use of electrical apparatus must be carefully specified and controlled to eliminate the risk of ignition where a dangerous concentration of vapour from highly flammable liquids may be present. Hazardous discharges of static electricity will be avoided by earthing electrical equipment, metal benches and storage containers.

Smoking and naked flames
Smoking must not be permitted in any paint refinish areas. Clear signs must be displayed at all access points (Figure 33). Never allow welding plant or blow lamps etc. within the vicinity of paint mixing or spraying areas.

In the event of fire, take suitable precautions against inhalation of smoke and products of combustion. *Do not* use jets of water on burning liquids.

Types of fire extinguisher
Red Water, used for wood, paper, textile, fabric and similar material (not for burning liquid or electrical).

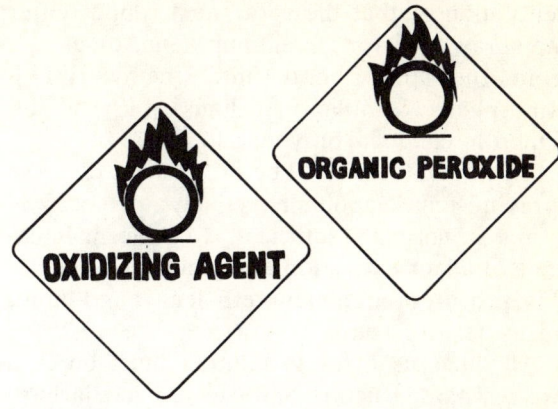

Figure 32 *Control against sources of ignition*

Figure 33 *NO SMOKING signs*

Black CO_2, for burning liquid and electrical.
Green BCF for burning liquid and electrical.
Cream/white Foam, for burning liquid (not electrical).
Blue Dry powder, for burning liquid and electrical.
Fire blanket Used for smothering.

The Factories Act 1961 (Equipment)

All plant, equipment and working conditions must comply with the statutory requirements as contained in the Factories Act 1961.

Recommendations for spraying operations (particularly isocyanate paints)
Spray operations should be confined to spray booths or enclosures under effective extraction/

ventilation so that the spray mist cannot enter the spray operator's breathing zone, or escape from the spray booth into the workshop atmosphere. Ventilation exhaust must be designed to conduct spray mist to a safe place.

Air flow velocity must be sufficient to disperse spray mist and vapours effectively; a velocity of 1 m/s is normally sufficient. Regular maintenance of all air extraction equipment is essential; this procures equipment efficiency and minimizes health hazards.

All ducts used for ventilation must be of a fire-resistant structure and cleaned regularly to prevent the accumulation of solid paint residues. Cellulose and oil-based synthetic paints should not be sprayed in a workroom with dry extraction because of interaction; the solid residues from cellulose are highly flammable and may be ignited by spontaneous heating of the solid residues from the oil-based paint. A water-washed extraction system must be used under these circumstances.

The design of equipment used to supply compressed air to operator breathing apparatus must conform to BS 4275: 1974. Care must be taken to ensure that the supply of air to the compressor is drawn from an uncontaminated source and that an efficient oil/water and fume filter is fitted. Where operators work within the booth an alarm system should be fitted to airline respiratory equipment to warn the user of diminished air pressure below a safe working level.

Paint mixing area/storeroom
The storeroom and workroom should be separate and constructed to an approved standard with stored paint not exceeding the maximum quantity licensed by local authority and conforming to the petroleum regulations. Ensure that there is adequate ventilation. A separate ventilation system may be necessary for the mixing room and for the paint store; this *must* meet the requirement of the Factory Inspectorate. An area or a cabinet specifically designed for spraying all test panels should also be available.

Environmental health and safety for refinish operators

Relevant Standards

1 BS 4275: 1974: *Recommendations for the Selection, Use and Maintenance of Respiratory Protection Equipment.*
2 BS 4667: Part 3, 1974: *Fresh air hose and compressed air line breathing apparatus.*
3 BS 2092: 1967: *Specification for Industrial Eye Protectors.* The Protection of Eyes Regulations 1974, Statutory Instruments 1974, no. 1681.
4 BS 2091: 1969: *Specification for Respirators for Protection Against Harmful Dusts and Gases.*

Materials

All paint materials must only be used in areas with excellent ventilation. Efficient air extraction units may be required in addition to the recommended respiratory precautions where large quantities of dust are generated during preparation with machine sanders. Employees *must not be* exposed to such atmospheres without a face mask that performs to the relevant standard for dust particle filtration. The face mask is a vital item in any paint shop and different types of material and atmospheric contamination require specific types of mask; always use the correct type.

Dry sanding operators require protection to the equivalent of a cartridge respirator conforming to BS 2091: 1969 fitted with type B dust protection; note that zinc chromate is particularly hazardous in the form of both sanding dust and spray mist.

Operators exposed to spray mist and working inside the spray booth must wear compressed airline breathing apparatus to BS 4667: Part 3, 1974, or other air-fed equipment which provides equivalent protection, both during the spraying process and until the spray mist has dissipated.

Most paint solvents will cause dizziness or loss of consciousness if inhaled at high atmospheric concentrations; prolonged and repeated exposure to these organic solvents can ultimately cause permanent damage to the liver and kidneys. Inhalation of spray dust and particles of dust generated from sanding will also cause lung damage; this must always be avoided.

Persons entering the spray booths or enclosures for short periods when spraying is taking place must be protected from inhaling the spray mist by wearing suitable respirators to BS 2091: 1969 with charcoal canisters (type CC, vapour/particulate). Care must be taken to ensure that filters are changed periodically and that facepieces are washed regularly with soapy water and a mild sterilizer.

Protective clothing

Complete protective clothing consists of a full facepiece airline breathing apparatus and eye protection, together with overalls made from an impermeable material. Splashes of solvents, paint activators and additives can cause severe eye damage; consequently, chemical-type goggles to BS 2092: 1967 are essential whenever there is a possibility of paint splashes entering the eyes, e.g. when opening tins, mixing, etc. These should also be worn as a protection against overspray. Eye protection must conform to The Protection of Eyes Regulations 1974, Statutory Instruments 1974, no. 1681.

Skin contact

Many paints and refinish chemicals will cause skin irritation after prolonged contact (Figure 34) and must be removed promptly, if necessary using a cleansing material. Paint solvent used for removing splashes may cause dermatitis, particularly when skin has been in contact with peroxide hardeners or acid catalysts that have a drying effect and often produce unnoticeable burns. All personnel who handle or use paints should wear overalls and gloves. A suitable barrier cream may help to protect exposed areas of skin.

Paint containing isocyanates

Precautions are necessary when spraying all paints but there are hazards peculiar to isocyanate synthetic paints.

Toxic effects

The initial effect of sufficiently high concentrations of isocyanates is general irritation of the lungs and respiratory system, causing dry throat, coughing, chest tightness and wheezing. Asthmatic attacks may also occur either immediately or some hours after exposure. Some operators may develop an allergy to isocyanates which will be evident by an asthmatic condition when exposed to even mild concentrations. Recovery is normally rapid when exposure ceases but repeated situations may result in a more permanent ailment.

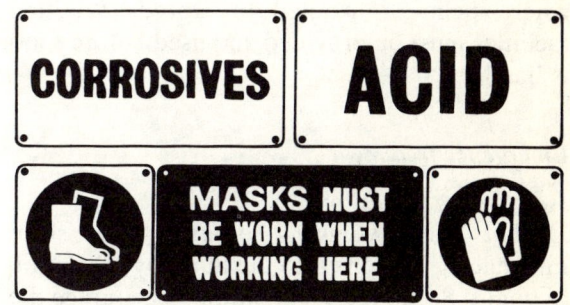

Figure 34 *Skin contact signs*

Figure 35 *Hand washing signs*

Personnel

Persons prone to asthma must avoid products containing isocyanates. Refinishers engaged in spraying products containing isocyanates are advised to have operators medically examined before employing them in this work. Anyone showing adverse symptoms must obtain medical advice immediately.

Isocyanates are also mild skin irritants and may cause dermatitis. Splashes of paint on the skin should be removed promptly with a cleansing material and then by washing the affected area with soap and water (Figure 35). Paint thinners or solvents should not be used.

Accidental ingestion of paint may cause severe irritation of the mouth, throat and stomach resulting in vomiting and pain. Drowsiness or loss of consciousness may result from excessive absorption of solvents.

Food should not be brought into, stored, prepared or consumed in the workroom or areas where paint is handled or stored. Washing facilities must be provided and used before food is handled.

Emergency treatment

Isocyanates (ingestion)
Do *not* induce vomiting because of the risk of introducing solvents into the lungs by inhalation of vomit. Seek medical attention. Encourage the patient to drink one or two cups of milk.

Solvent vapour
Move into fresh air, keeping the patient warm and at rest.

Eye contact
Splashes of paint will irritate the eyes and should be treated promptly by washing with clean water. Obtain medical attention immediately.

General environmental health and safety

Spillage
Accidental spillages should be absorbed with sand or an inert material. Disposal of waste materials collected in this manner must comply with the Deposit of Poisonous Waste Act 1972. *Do not allow* the materials to enter any drains or water courses.

Cleanliness and tidiness
A clean and tidy work area is necessary both to avoid accidents and to produce good quality work. Keep floors dust free. Spilt liquids and grease must be wiped up immediately. Disposal of paint residues, contaminated wastes, etc. must comply with the requirements of the Deposit of Poisonous Waste Act. Empty tins, parts and equipment must not be left in areas where they will obstruct movement of personnel.

3 Hand tools and panel beating techniques

Damage assessment

Accidental damage and subsequent repair ranges over a broad front, from the complete 'write-off' to the low-speed 'parking' incidents. The age of the car, its condition and its market value will determine the insurance approval for repair. Major accident repair work involves the use of the manufacturer's replacement parts and the reconstruction of the body on an approved jig.

Repair work without reconstruction on a jig and without welding in replacement parts in full, or as a section, is possible when wheel alignment and chassis alignment are proved to be within limits on checking before work is put in hand.

Restoration of damaged body panels calls for a high degree of skill, not only in the use of tools but also in the ability to assess the damage, to determine the cause and extent of damage and also the sequence in which damage occurred to various areas.

It will be appreciated that in crash work there can be two types of damage – direct damage in the area in actual contact with the object causing damage, and indirect damage in the areas surrounding the direct damage. These areas of indirect damage may – in some instances – be a considerable distance away from the point of actual contact in a crash. For instance, imagine a car somersaulting and making impact on the ground with the front edge of the roof. Indirect damage can be caused at the rear end of the roof by the forward action of the car pushing the roof back; sideways motion could also push out the roof to either side; and finally on rolling over the impact on the top could give a downward motion. These are three examples of indirect damage from one point of direct damage.

Apart from scratches and panels punctured or torn, there are three other types of damage which have to be dealt with. These are ridges or peaks formed by bending or buckling of the panel beyond its elastic limit; sharp channels or depressions (the opposite of ridges); and also hollows or contours where the metal has been distorted within the limits of elasticity. The last type of damage is prevented from returning to its original shape owing to the action of ridges or channels, or a combination of both, in the surrounding area of damage.

If the stresses formed in the ridges on impact can be released, then the area of hollow will return to the original shape. It will be appreciated that, in all but minor crashes, several ridges or channels will be formed and it is necessary that all be relieved of their respective stresses to enable the body sections to return to their required shape. It is important that these corrections be made in the right sequence as otherwise additional damage will be caused to the panels. Work must be carried out directly opposite to and in the reverse order of the forces causing the damage. In other words, the last ridge formed in the crash should be removed first, working through the damaged areas until the first point of impact is the last to receive attention. We will use as an example the car mentioned as somersaulting. The first indirect damage was pushing back the roof and it will be realized that to attempt to correct this first will result in exaggeration of the downward strain or the pushing further out of the sides. In such a case, first the crushed-in top should be corrected, followed by correction of the sideways motion, and finally the bringing forward from the rear end.

An experienced light vehicle body repairer will very quickly assess the cause of the damage and follow the sequence from the point of direct damage through the various stages of indirect damages, but until this experience has been gained it is essential that all details be obtained of the collision so that it is possible on locating the point of direct damage to determine the direction from which the damaging force came and follow through the course of resulting indirect damage.

Great help can be gained with this work, particularly where reinforced sections are damaged, by the use of body jacks which when applied correctly exert great pressure at the required point – a pressure which, unlike heavy hammering, can be controlled, whether applied by mechanical or hydraulically operated units. Modern techniques have demanded specialized tools and a wide range designed to give full assistance to the body repairer is now offered, correctly styled to meet the varying contours of present-day bodies, balanced for ease of use and correctly forged and heat-treated for long service.

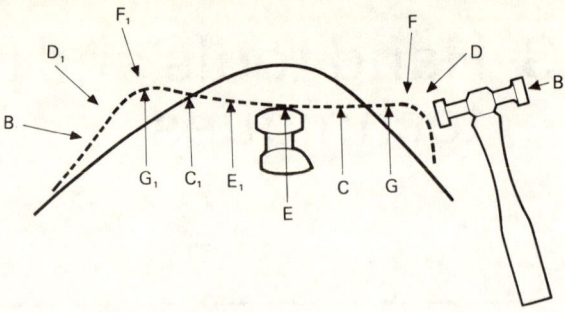

Figure 36 *Corrective sequence*

also ensure a reduction in wear and tear on the dollies etc. to be employed.

The first step in the restoration sequence is to unfold the valleys, V channels and buckles in the indirect damaged area as gently as possible without further stretching or reforming as they are brought up to something like their original position and contour.

They have not been stretched beyond the elastic limit but are held by the ridges at the outer edges of the damaged area. The reshaping should continue with alternative working on the ridges and the low areas as the metal is slowly raised up into line with the surrounding sheet. Figure 36 shows a double-end hand dolly and suitable beater in such a sequence.

Panel preparation

Before any repair procedure can be carried out on a given panel, the inner and outer surfaces must be thoroughly cleaned; deadeners, underseal and other foreign matter that might interfere with the application of corrective forces must be removed.

Most of the anti-drumming and other undercoatings used today can be removed by scraper or putty knife after softening by the application of heat to the outside of the panel (a large-tipped welding torch with a mild reducing flame should be used). The outside of the panel should be washed down with clean water and any traces of oil, road tar or asphalt removed with a solvent soaked rag. This panel preparation will make hand tool straightening more effective, and will

Beaters and beating

The amount of corrective force required will depend on the gauge of metal to be straightened and the extent of the damage. The repairer can control the intensity of the blows to be delivered; body and arm movement can be varied from light blows, controlled by finger movement or wrist movement, to blows that can be developed from the elbow or shoulder or even to two-handed blows involving movement of the body from the waist. The actual force delivered to the metal surface will also depend on the weight of the beater used, the weight and size of the dolly or spoon and indeed the area of actual contact with the panel. Too great a force between the beater and dolly can stretch the metal locally, moving it outwards or inwards, as the metal cannot move sideways to be absorbed within the panel. This action will involve further correction by shrinking.

The operation of beating metal is one of the highest skills in the whole business of vehicle body repairing. The ability to control the intensity of the blows delivered, together with the knowledge of how and where to direct the blows, anticipating the result to be expected, is the mark of the expert. Expertise can only be achieved by continued practice, once the basic requirements are understood.

To develop striking power using the fingers only, hold the beater as at position 1 in Figure 37; note that the forearm is held at an angle of about 20 degrees to the horizontal and remains in that position throughout. The beater is held loosely, the shaft resting against the base of the thumb and same distance from the heel of the palm, the fingers hooked over the shaft as shown. Now by closing the fingers to grasp the shaft, the head of the beater is thrown forward by the strength of the fingers to position 2. This method of panel beating is most useful when working beneath surfaces when the beater or the working area may be out of sight. When trying this method of beating for the first time, only feeble blows may be possible; strength in the fingers can only be developed by practice. Throw the head of the beater at the palm of the other hand, without movement of the wrist or elbow; alternatively, use a block of soft wood to show whether the blows are being delivered squarely.

Striking with the combined action of the fingers and wrist can be developed by following through with added wrist action (Figure 38). The movement from 1 to 2 is exactly the same as in Figure 37 but a follow-through action of the wrist will cause the beater to make contact with the panel at position 3. Some difficulty may be experienced at the initial attempt to strike with this co-ordination of movement; practice is again essential.

Heavier blows, for roughing-out purposes, become more in line with the natural movement of the elbow and shoulder, as in the use of ordinary hammers.

In early attempts at panel beating, some difficulty may be experienced in aiming the beater at the panel to strike squarely on to a dolly below. Place a small dolly below the panel and tap the top surface lightly with the beater. Find the high surface of the dolly by checking for maximum rebound. Once this can be achieved, move the dolly about following the movement with the beater to develop co-ordination until fairly rapid action is possible.

Beaters should be well balanced, i.e. the length of the shaft should give a feel of balance,

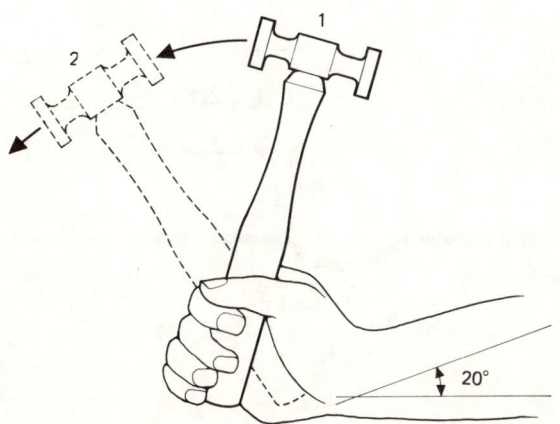

Figure 37 *Beating using fingers*

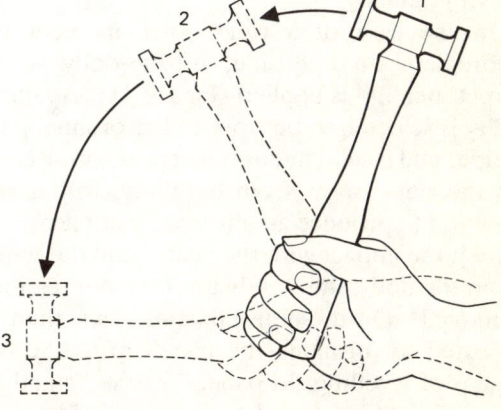

Figure 38 *Follow-through wrist action*

when the tool is held in the hand, at a point about three-quarters of the shaft length from the head. The handle should not be gripped tightly as this can cause fatigue in the arm muscles when beating over extended periods. When beating on metal the blows should land squarely on the surface. In all dinging operations the beater should travel in a circular path with rhythmic action of some 100 to 120 blows per minute (Figure 39). In this manner the metal receives a sort of sliding or glancing blow resulting in but a small area of contact with the surface.

To level out a panel, the beater should be moved about in regular rows, striking the metal at intervals of about 10 mm with light blows until levelling is completed. Beaters of sufficient size and weight such as roughing out and bumping are often used alone, or in conjunction with a piece of hardwood to raise the elastic areas of the metal. The non-elastic outer ridges are then reshaped by 'spring-beating' or on- and off-the-dolly techniques.

On-the-dolly or 'direct' beating is shown in Figure 40. The dolly is selected so that its contour, to be held under the ridge is near to the original shape of the panel at this spot. Beating is then directed at the peak of the ridge, commencing with light blows, increasing in intensity to a level sufficient to push the ridge back. Work along the ridge from end to end in a progressive manner, i.e. do not flatten the ridge completely in a localized area, but take it all down gradually.

In the case of a ridge with an associated depression on one side, off-the-dolly or 'indirect' beating is applied (Figure 41). Again the dolly is selected to be close to the original panel shape, and is held under the depression. Beating on the ridge, away from the dolly, will cause a reaction to produce an alternating impact on the panel; the impact with the beater and the impact from the dolly alternately until the depression is removed. On-the-dolly beating can then be resorted to, bringing the panel up to the final stages of levelling. In panel finishing, small low areas should be raised by using the side of the round face of the beater initially. The surface is

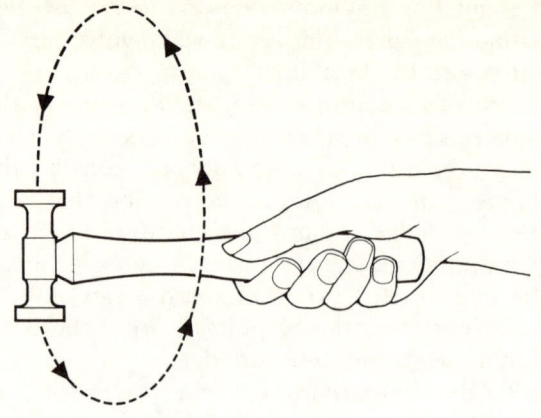

Figure 39 *Dinging action*

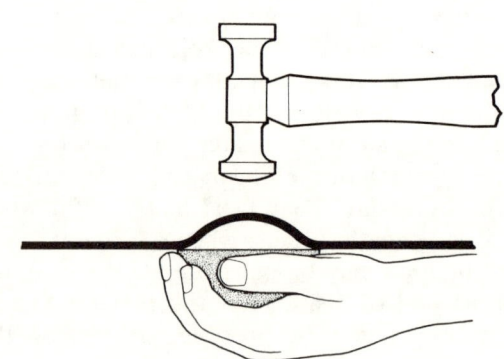

Figure 40 *On-the-dolly technique*

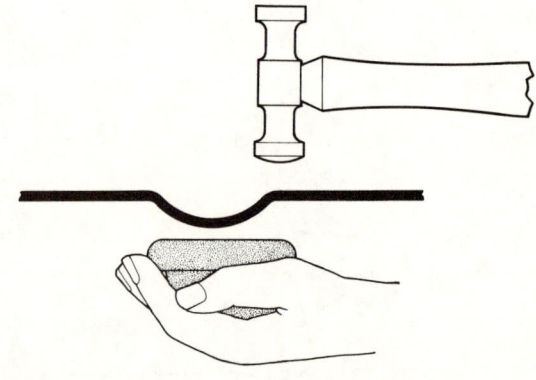

Figure 41 *Off-the-dolly technique*

then checked with the body file, to highlight any remaining low spots. Each spot should be raised individually with the pick (Figure 42). Care must be taken not to strike the low spots too hard otherwise they will become rough and 'pimply'; the metal will be stretched and subsequent shrinking may be necessary. Use of the body file on the metal in a pimply condition can cut holes in the panel if the low spots have not been picked up smoothly. This finishing operation requires a considerable amount of patience, a great deal of skill and a good 'eye' if it is to be carried out efficiently.

When a panel is subjected to extra-heavy stresses, with a collapse or distortion of the surrounding framework or inner construction, roughing out will be beyond the capability of hand tools and hydraulic equipment will be used. Once the initial straightening is accomplished, hand tools should be used to level out the ridges and raise the depressions before the hydraulic load is released.

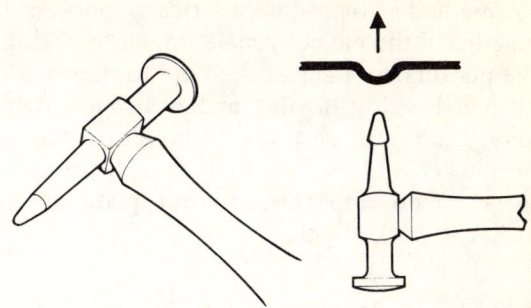

Figure 42 *Raising low spots*

Mallets

Mallet heads are usually made of hardwood fitted to ash handles, to form a complete mallet. Standard round mallets have cylindrical heads. Bossing mallets combine the hemisphere and cone and are sometimes referred to as pear-shaped mallets. Both patterns are available in other materials, such as rubber, rawhide and the softer metals as in copper-faced mallets. Rubber mallets with interchangeable screw-on heads are ideal for use on aluminium and may be used in the manufacture of panels in sheet steel.

Hollowing is a process of thinning metal from the centre of a given blank to produce a double-curvature panel (Figure 43). This can be accomplished by the use of a bossing mallet and a sandbag. Beating of the metal commences in the centre of the plate, working outwards in increasing circles until the required curvature is obtained. Uniformity in the intensity of the blows is important to produce a regular shape.

Raising is an opposite effect, in which the metal is thickened up at the edges (Figure 44).

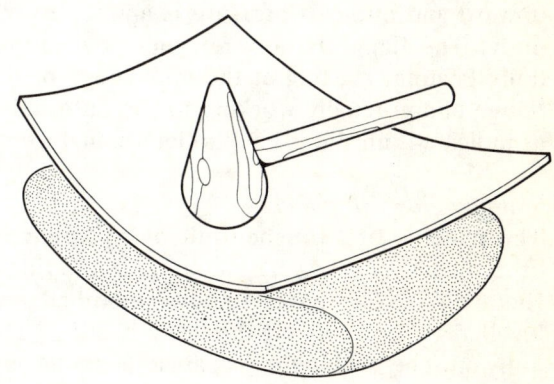

Figure 43 *Hollowing*

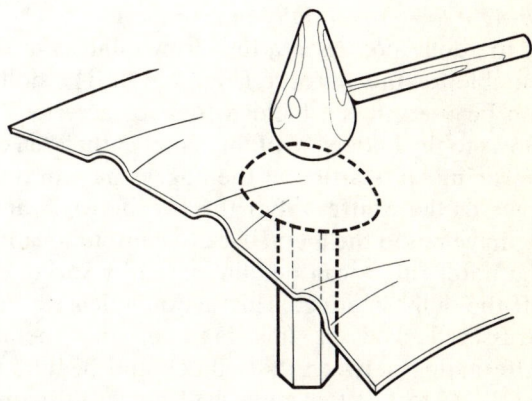

Figure 44 *Raising*

The method is to produce a series of 'puckers' in the edge of the panel by malleting or 'blocking'. The puckers are then worked out to the edge of the panel with a bossing mallet and a suitable dolly. Care must be taken to hold the metal in the pucker and use the mallet to disperse the excess metal into the adjacent plate as the pucker is worked out.

Applications of beater and dolly

Reshaping a flange
The utility dolly is placed in the damaged flange using the edge most suitable to the shape and size of the original flange (Figure 45). An upward and outward pressure is applied to the dolly. The flange is now reformed by on-the-dolly beating, starting at the inner edge of the flange and gradually working to the outer edge as indicated, until it is back to its original form.

Shaping crowned panels
The panel is first roughed out by hydraulic or other means of releasing the metal in the ridges (Figure 46). The panel is now smoothed and levelled by on- and off-the-dolly beating. The dolly must be held tight up against the metal as it is moved backwards and forwards across the crown during the beating. The choice of the dolly will depend to some extent on the shape of the original crown. The dome dolly is illustrated in this situation.

Using a low-crown dolly in reshaping
A toe dolly possesses a low crown and is most suitable for this purpose (Figure 47). The dolly can be used in the hand to deliver a series of blows to the underside of the panel in the area of elastic metal, starting at the edges and working towards the centre. After the two blows A and A_1 have raised the metal back to contour locally, the outer ridges can be eliminated by spring or off-the-dolly beating. This action releases any stresses locked in this area of the metal. Alternating between dolly block and beater, B to B_1, C to C_1, the metal will be raised to its original shape.

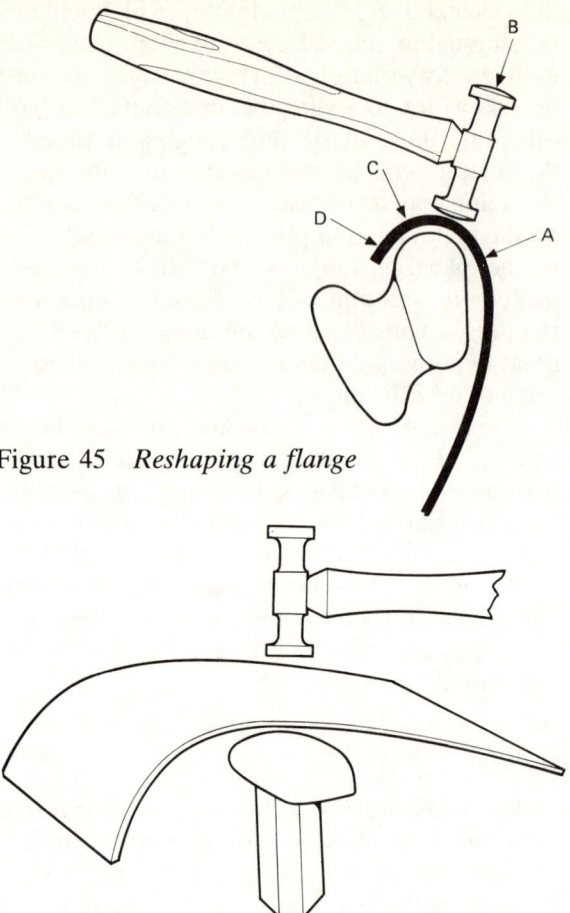

Figure 45 *Reshaping a flange*

Figure 46 *Shaping crown panels*

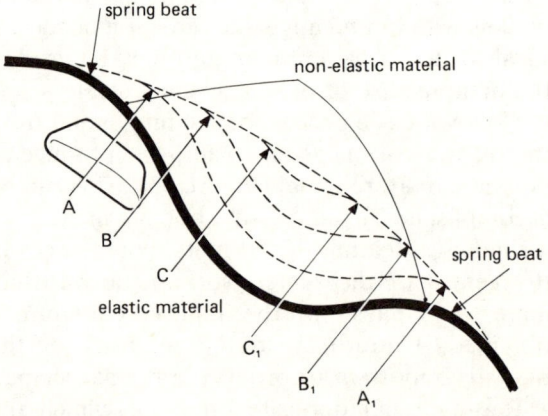

Figure 47 *Use of low-crown dolly*

Raising a low spot with a curved dolly

Holding the dolly lightly against the metal, use the off-the-dolly technique beating at A and B to lower the ridges and at the same time raise the metal because of the light support given by the dolly (Figure 48). If the low spot is still lower than the surrounding sheet metal, strike the underside of the panel, one, two or a series of blows, until the panel is raised to its proper level. The dolly can now be turned on to its face so that on- or off-the-dolly beating can continue until straightening is completed.

Reshaping a roof panel with a curved dolly

The edge of the curved dolly can be used just above the roof rail to restore the shape of the roof (Figure 49). Repeated blows with the wide edge over the damaged area will raise the metal as a roughing-out process. The low-crown face or other suitable area of the dolly can now be used as in on- or off-the-dolly beating to make the restoration complete.

To reform or manufacture a bead or moulding

The shape of the final bead will determine the best dolly to use; in Figure 50 the angle dolly is used. If the bead is to be manufactured on a replacement panel or patch plate, its size and shape must be clearly marked along the panel or patch plate with three lines to show the crown and other edges of the bead. The dolly is held lightly under the centre line of the marked-out metal and it is beaten until it is slightly stretched. The outer edges and the ends of the bead are then formed by off-the-dolly beating until the metal is stretched sufficiently to hold the bead in position.

To restore the shape of a bead

As in Figure 50, the angle dolly is placed in the damaged bead and pressed firmly against the inner surface. The panel is now beaten on each side of the bead as indicated by the arrows, shaping the metal around the face of the dolly. This action continues along the entire length of the bead until it is of proper shape.

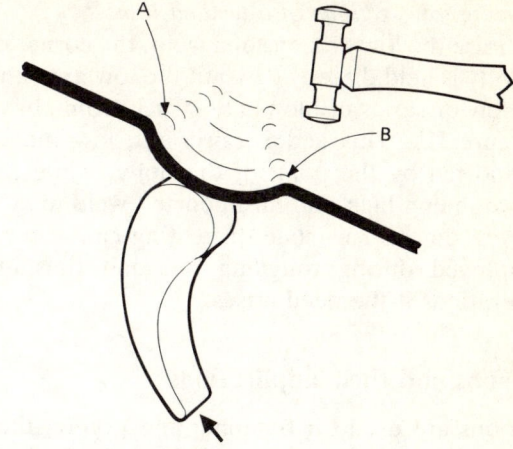

Figure 48 *Raising a low spot*

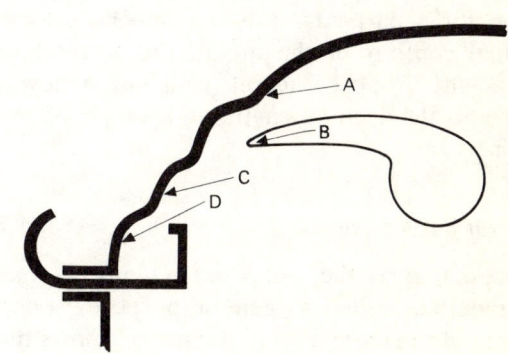

Figure 49 *Reshaping roof panel*

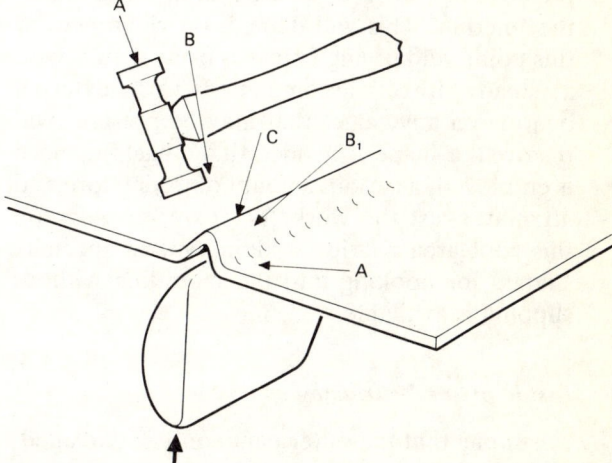

Figure 50 *Reforming a bead*

Raising low sections of a welded joint

To raise the low sections of a weld, the corner of a dolly is held directly beneath the low area and a blow or series of blows are struck from above (Figure 51). This action raises the low metal, supported by the point of the dolly, while the surrounding high and unsupported weld area is driven down. This style of beating can also be employed during roughing out and finishing operations if the need arises.

Spoons and their applications

Spoons are used for bumping and prying; they are also used in place of dollies when direct access to the rear of the panel is obstructed by the internal frame structure. The choice of the spoon for a particular job will depend on the original contour of the metal, the amount of access, the proposed action (roughing or levelling) and the general shape and length of the spoon.

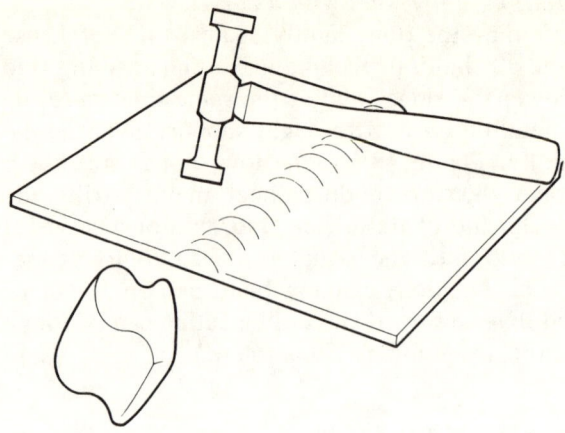

Figure 51 *Raising low sections*

General-purpose spoon

The reshaping of the roof panel in Figure 49 may be possible using a general-purpose spoon instead of the curved dolly. Figure 52 shows the application. The use of a backing piece prevents local damage to the cant rail and reduces the pressure in this area by distributing the force at the fulcrum. This will depend on clearances at this point, and prying upwards from A to B with a steady force, accompanied with external beating on any ridges that may be present, will restore the shape. Provided that a backing piece is employed, a certain amount of prying fore and aft may assist the work. For exterior work on this roof area a drip moulding spoon specially shaped for hooking into the moulding without slipping is available.

Inside pry and surfacing spoon

Assuming that the outer panel only is damaged, making the repair an economical proposition, the car door provides an example of a double-

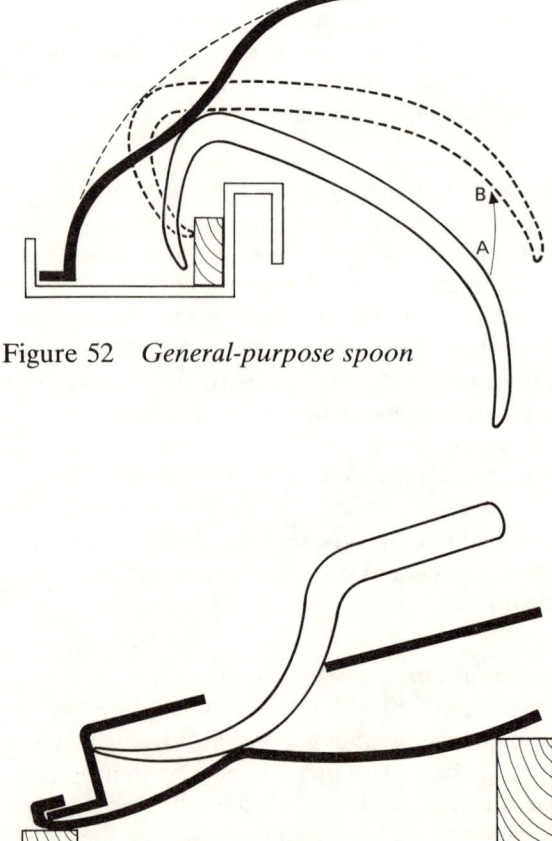

Figure 52 *General-purpose spoon*

Figure 53 *Inside pry spoon*

skin structure to illustrate the application of body spoons (Figure 53). When the door is stripped of interior trim and window glass it should be placed panel down towards the floor or bench, resting on two pieces of wood. This prevents the panel scraping the floor and gives space for the panel to move or spring as force is exerted on the spoon. After roughing out the spoon can be reversed so that on- or off-the-spoon beating can be used to complete the straightening procedure. Access for this spoon will depend on the pattern of the piercing through the inner structure. If direct access is not possible a long-reach dolly should be used. Alternatively, circular holes may be cut in the inner panel for access and flanged after repair to be covered by the trim on reassembly.

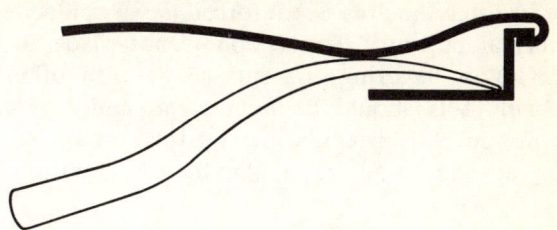

Figure 54 *Heavy-duty pry spoon*

Heavy-duty pry spoon

This spoon (Figure 54) can be used for separating outer panels from the inner frame structure when they have been damaged and squeezed together. The spoon can be driven between the plates prying sideways or up and down until the desired amount of separation is achieved. The blade can then be used as a dolly to dress out the outer panel and the inner structure, if required. The blade is reduced to a very thin section which can be used for opening door panel flanges or breaking spot welded joints that have been previously drilled through the panel or section to be discarded.

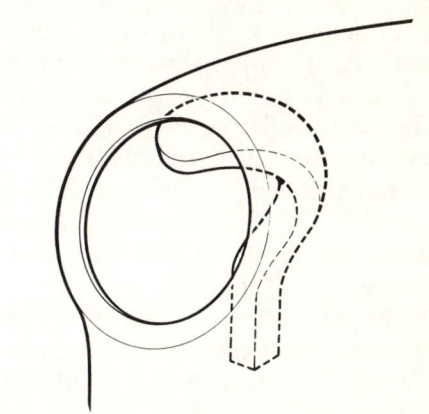

Figure 55 *High-crown spoon*

High-crown spoon

The high-crown spoon (Figure 55) with its broad working surface and high crown is an ideal tool as a dolly or as a spoon for work in confined areas such as headlamp housings and high-crown sections of the body.

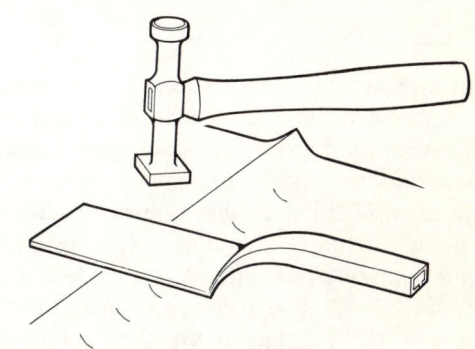

Figure 56 *Spring-beating spoon*

Spring-beating spoon

This is a light pressed-steel spoon (Figure 56) designed specially for spring-beating on ridges. The spoon is placed directly on the ridge and sharp blows with a beater are delivered to the back of the spoon, spreading the force over a large area. In this manner marking of the panel is prevented and the damage is corrected in many cases without injury to the paintwork. The intensity of the blows should be closely controlled so that the area is not forced down below its normal position. This spoon is not made for prying or levering. Its surface as with other panel tools should be kept clean and highly polished. Any irregularities on the surface will be reproduced on the panel in the reverse order.

Bumping blades

Bumping blades (Figure 57) are used for slapping out dents, with or without the backing support of a dolly. For slight dents or a' wavy surface a dolly is not required; the bumping blade should be applied so that glancing blows are received by the panel. The blade serrations hold the metal within the area of contact to avoid stretching. Use should be limited to slight or moderate damage as this tool is not intended to take on the role of the body spoon or beater and dolly. Experience will indicate when to use the bumping blade and it should be tried out as a semi-finishing tool before the use of a body file for practice in use.

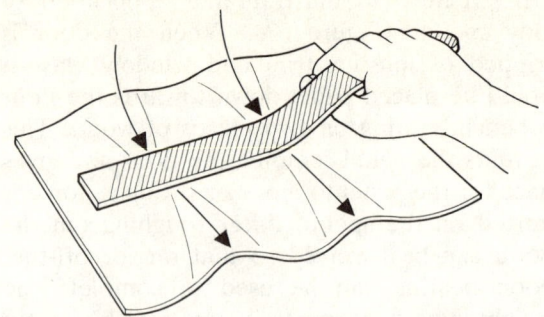

Figure 57 *Bumping blade*

Specialized tools

Long-reach dolly

A compromise between the dolly and spoon has been reached in the long-reach dolly where the head is separated from the shaft to gain access not possible with dolly or spoon. Figure 58 shows an application of the long-reach dolly to an outer door panel. A small hole is drilled in the inner panel to accommodate the shaft while the head of the dolly is passed through a normal access hole to be fitted to the shaft within the door structure. A special button dolly will enter through a 32 mm hole, a useful tool for door pillars and other box sections.

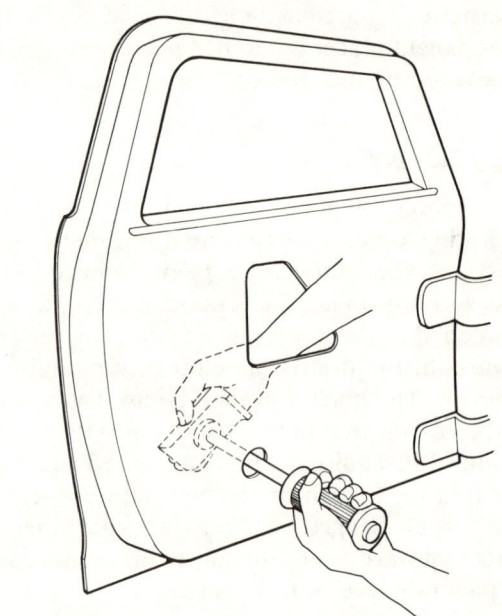

Figure 58 *Long-reach dolly*

Panel puller

The panel puller (Figure 59) adopts the principle of the slide-hammer for its pulling action. Removal of the interior trim and other lining material can also be avoided in many cases.

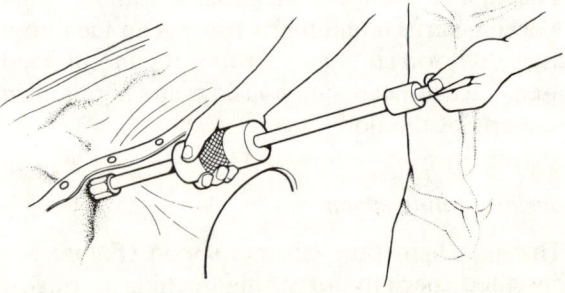

Figure 59 *Panel puller*

The standard practice is to punch or drill a 3 mm diameter hole in the deepest part of the indentation; screw in the self-tapping screw until a firm purchase is obtained. Now by pulling on the cross-bar of the tool and impacting the sliding weight against the forged stop on the shaft, dents can be effectively removed. The holes made for the self-tapping screws can be filled by brazing or welding on completion of the straightening operation.

As an alternative method a suitable adaptor could be made to replace the self-tapping screw and hexagonal sleeve of the puller. The paint should be removed from the panel and the metal effectively cleaned and tinned. The made-up adaptor is soldered in position and the panel puller shaft screwed home. The puller is now operated as described.

If further pulling is required within the damaged zone the made-up adaptor can be moved provided that tinning and soldering is carried out as before. On completion the damaged area can be solder filled as the first stage of this operation has been carried out. This method of repair will obviate the need to fill the holes made by the tapping screws.

Panel finishing

Shrinking

In the stages of panel finishing some slightly stretched areas of metal may be encountered. The repairer determines the exact location, size and shape of the area by 'hand-feeling' (running the palm of the hand over the surface), by 'eye' or, in the case of a large flat area such as a door panel, by the use of a straight edge. To increase sensitivity when hand-feeling a lightweight cotton glove could be used to feel the panel. On a paint finished car body a spray of light oil and a strong light source can be used to show slight irregularities in the surface. Once the stretched area is located the treatment will depend on the amount of stretch involved.

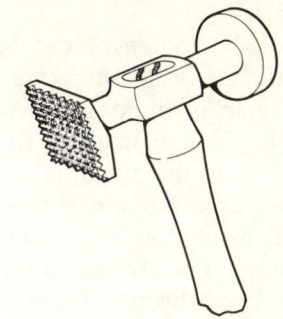

Figure 60 *Shrinking beater*

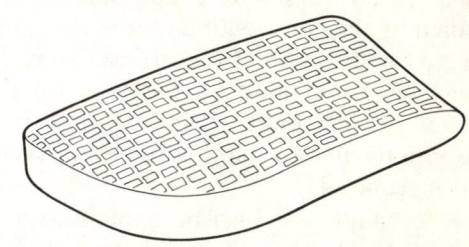

Figure 61 *Grid dolly*

Shrinking beaters

For slightly stretched areas shrinking beaters may be used (Figure 60). The square-faced end of the beater is cross-milled, and in using this type of beater small amounts of metal are forced into the spaces created by the cross-milling. This type of beater can be used on aluminium in a cold condition, but working on sheet steel will be speeded if heat is applied to the area. Whenever possible such beating should be carried out on the inside of the panel to minimize resurfacing work prior to painting.

The grid dolly

A special dolly available to facilitate shrinking work is the grid dolly which has a large crowned grid face on the upper surface (Figure 61). The base is a shallow crossed face for normal finishing work. A combination of shrinking beater and grid dolly, combined with the application of heat, provides quick and effective shrinking.

Shrinking dolly

Panels subjected to intense stretching can be tightened by the use of the shrinking dolly which has a groove machined along one face (Figure 62). A shrinking beater, made to fit the groove less the thickness of the metal, is also required.

The method used to cold shrink the metal involves the construction of ribs on the underside of the panel. The grooved dolly is held on the reverse side of the panel and the metal is driven down into the groove by beating. The position of the ribs produced and their length will be determined by the area of the stretched metal. A stretch in one direction only could be remedied by the construction of one rib at right angles to the direction of the stretch. To reduce a bulge an X rib may be the best solution. However, the job could be checked for 'panting' at the introduction of the first rib, and a second added if required.

This technique can also be applied to welds (Figure 63). The weld (a) is forced into the groove of the dolly by beating to rib the panel as at (b). The strength of the weld has not been reduced by sanding or grinding and the stretch usually present in a welded joint is removed. The ribs produced to remove the 'pant' or set the weld below the panel surface can be built up by means of a suitable filler as at (c).

Hot shrinking

Hot shrinking is carried out with the aid of an oxy-acetylene torch; the size of the tip should be as required to weld that gauge of metal with the flame adjusted to the neutral setting. Care should be taken not to hold the flame too near the metal as this will cause overheating or burn a hole through the panel. Determine the highest point of the stretched area and apply the flame until the metal is at a cherry-red heat. The size of the spot should be about 12 to 15 mm in diameter. Excess metal can be worked towards the spot by light use of a shrinking beater which avoids any stretching of the metal which might result from use of a regular beater. As the spot is heated the metal will rise to form a bulge above

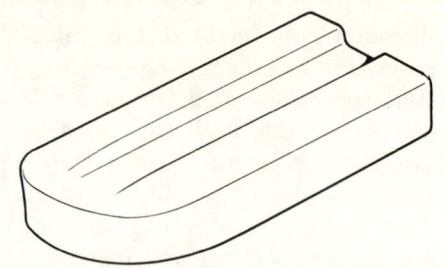

Figure 62 *Shrinking dolly*

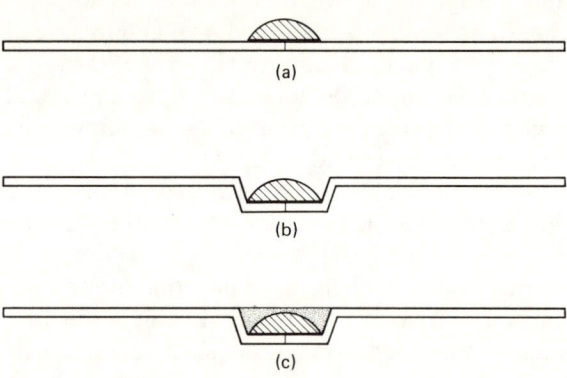

Figure 63 *Weld shrinking*

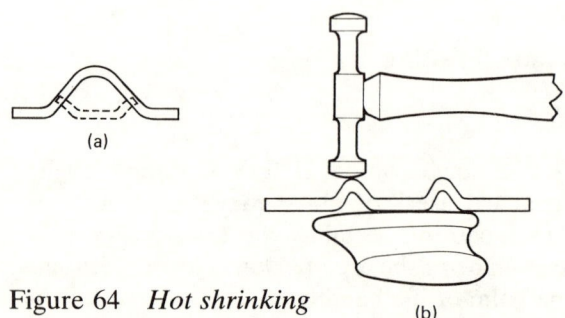

Figure 64 *Hot shrinking*

the surrounding metal. The bulge is then given a number of sharp accurately placed blows to cause it to collapse, as in Figure 64(a).

A dolly or body spoon is then held directly underneath the bulge with slight upward pressure. A dinging beater is now used to smooth out the ridge produced in the first beating operation.

The blows should be delivered directly on the ridge with a slapping motion towards the centre of the heated spot, as in Figure 64(b). The metal in the heated condition is soft, and so beating should not be excessive.

Once the heated spot turns black it should be quenched with a water-soaked rag or sponge. This causes the metal to contract quickly, tightening up the metal still further. The metal should not be quenched before it turns black, otherwise it will crystallize and harden to make panel finishing difficult.

For shrinking operations use a flat dolly on low-crown panels and a low-crown dolly or spoon on high-crown panels; action should be taken as quickly as possible before the metal loses its reddish tint and turns black. To prevent heat transfer to the panel, wet rag or cloth can be applied as a pancake to the area, spreading it outwards from the centre until the area to be shrunk can be seen and is of sufficient size for the operation to be carried out.

If the area to be shrunk is large, then shrinking operations should be repeated about the area, leaving a reasonable space of metal between each shrink. This method is known as sequence shrinking.

The underside of 'hot-shrunk' panels are prone to rapid rusting and special care should be exercised to see that they are primed and undercoated as soon as possible after repair.

Metal finishing

Filing

Filing is a means of metal finishing a damaged panel prior to sanding operations for paint spraying. File (or body) blades are made in flat or half-round sections for mounting on wooden holders to give a firm and rigid backing to each of the preset shapes, flat, half-circle, special curved or radiused blade (Figure 65). The standard blade of 356 mm and a mini-blade of 203 mm are available in a variety of cuts from 6.5 teeth per 25 mm to 17 teeth per 25 mm, depending on the material to be cut. One of the

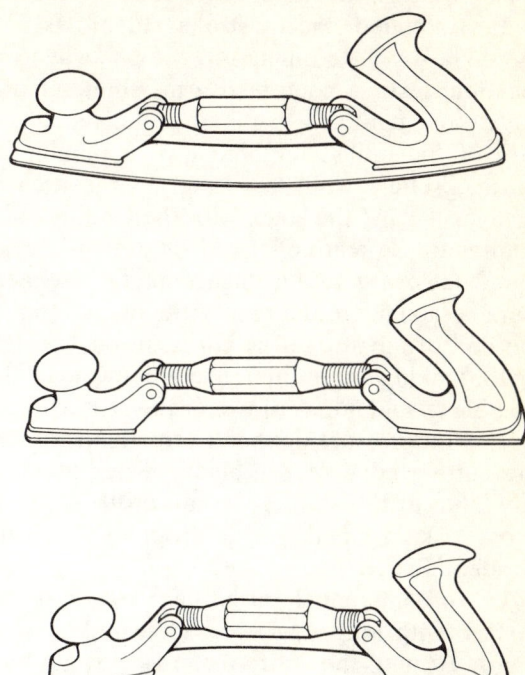

Figure 65   *Adjustable file blades*

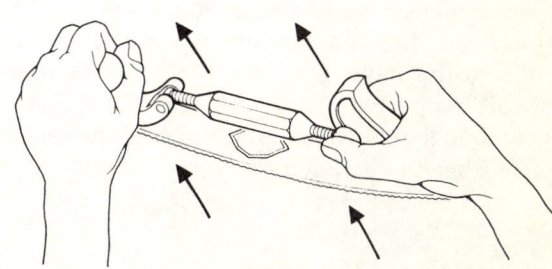

Figure 66   *File application*

most versatile types of holder is adjustable (concave to convex) over a length of 356 mm.

In use the blade should be adjusted so that it does not quite match the contour to be filed. Some clearance at each end of 1.5 to 3 mm is desirable. The file should be applied with long straight even strokes holding the tool with both hands so that it makes an angle of 25–30 degrees to the intended line of stroke (Figure 66). It should be pushed straight, directing it away from the body with a combined arm, shoulder and body movement. This action enables the curvature of the milled teeth to give a planing or shearing effect while covering a large area at each stroke. At the start, with the leading hand dominant, the teeth of the blade are cutting; as the stroke continues the engagement of the teeth works gradually to the rear of the file, so that at the end of the stroke the cut is on the last few teeth, making use of the full length of the blade. The file should not be drawn backwards over the work with pressure applied as this tends to dull the cutting edge of the blade. After use it is advisable to release the tension on the back to avoid risk of breakage, if dropped or struck accidentally.

To finish a panel push the file across the surface with long strokes as indicated at A in Figure 67 until the entire panel is showing bare metal. Note the angle of the file in relation to the direction of the stroke. Then file over the crown towards the centre of the panel as shown at B and C. This action will show up high and low spots for correction and draw the metal away from the crown to smooth the contour. The outer edges of the panel should now be filed from end to end to make sure that all irregularities have been corrected. Do not use the file to excess – it is not to be used as a correction tool but as an indicator to show where further beating is required.

Filling

Catalysed stoppers, such as polyester resins and a catalyst known as a 'two-pack' material, have become a standby for the repairer in the process of panel finishing as they can be applied directly to clean and scuffed bare metal. It should however be emphasized that panel beating should be of a reasonably high order so that filling by such means uses minimal quantities of surfacing only. The introduction of these filling materials brought with it other problems such as

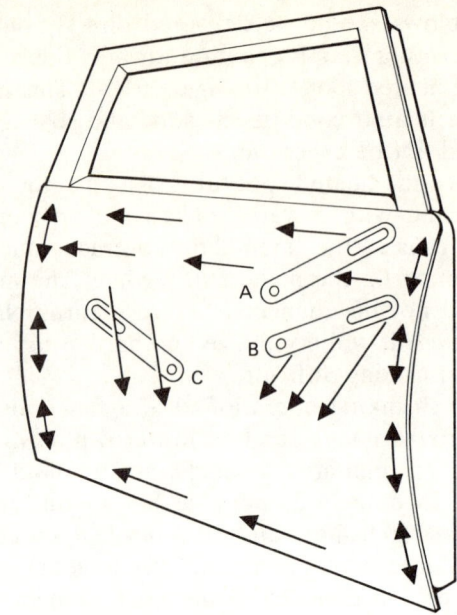

Figure 67 *Filing a door panel*

'clogging' of the normal metal finishing blade. A special 'plasticut' body blade with 6.5 teeth per 25 mm has been developed for use on these materials. The assumed loss of time in changing blades to the correct cutting pitch is offset by the improved speeds of cut when correct tools are used.

Care of tools

The pride of the artisan is the set of tools he possesses. Care of tools throughout their working life is important. All bright surfaces should be kept clean and free from scourers and blemishes that could be transferred to the body panels. Storage of tools when not in use is also important, but this will depend on the working conditions in the body shop. Wall boards with the necessary clips and tool silhouettes can be purchased with a specified set of tools. This is ideal for general usage from a central store as missing tools can be quickly identified. The same set of tools is also available in metal boxes which is most useful when each worker maintains his own tool kit. Tools can, of course, be purchased separately, enabling the body repairer to replace a damaged item to build up his own tool kit or extend an existing set to cope with new shapes in body styling. It is not good practice to shorten existing beater shafts if cracked or broken as this will affect the balance of the tool. Replacement shafts are available for all beaters.

It would be impossible here to consider all the basic principles in the use of hand tools for body repair work. Each job has to be treated on its own merits. All jobs vary in the type and extent of damage; the repairer never finishes learning about his job, and has to rely on his experience, versatility and ingenuity at all times. He has to know the cause of the damage and its result so that a correct assessment can be made, and he has to plan the correct procedure and method required to give satisfactory results.

4 Welding processes

Historical background

The history of modern metal joining began in the late nineteenth century with three discoveries. This is not to say that metal joining had not been carried out before then. In the Bronze Age, components had been joined by a casting-in process. In the Iron Age it was found that pieces of wrought iron could be joined by heating and hammering them together, and thus forge welding was developed.

In 1885 Bernados invented a process in which an electric arc was used to melt the edges of two pieces of metal and thereby join them together and hence began the development of a range of arc welding processes. About the same time Elihu Thompson found that when a heavy current was passed through two pieces of metal in contact they became joined together and this observation initiated the resistance welding processes. Also in the 1890s a famous French chemist, Le Chatelier, realized that if oxygen and acetylene could be burned together, the temperature of the resulting flame would be higher than that of any other flame known at that time and would be capable of melting steel. This led to the development of oxy-acetylene welding, which because of cheapness and versatility is widely used for jobbing work and limited production welding. The development of welding processes has followed a fairly coherent pattern. Soldering and brazing have evolved in response to a number of requirements and over a longer period of time, and present-day practice is determined by the properties of the available filler metals and the characteristics of heating sources.

Oxy-acetylene welding

Gas welding is probably the most familiar of all welding processes being used in every garage and workshop. A welding torch burning oxygen and acetylene produces a high-temperature flame to melt the metal and form a fusion weld, additional metal being added separately by dipping a filler rod into the weld pool. With suitable fluxes all of the common metals and alloys can be welded. In addition, cast irons, steels and copper are often joined with a 60/40 brass filler rod in the process known as bronze or braze welding, which because of the lower melting point involves less distortion. Another allied process is surfacing, in which layers of alloy steel, tungsten carbide/cobalt mixtures or bronzes are deposited on surfaces for wear or corrosion resistance. Only a minimum amount of inexpensive equipment is required, so that gas welding is highly suitable for jobbing and light production; it is comparatively portable and operates independently of any power supply.

The heat required to form the weld pool is imparted to the plate by collision of the high-energy molecules in the gas flame. This is less efficient and the heat is distributed over a larger area than when heating with an arc; there is more distortion, and it is more difficult to weld thick sections than with the arc. For the joining of thin sections gas welding is superior to shielded metal arc welding as more delicate control of the heat input is possible, and the arc forces that tend to form holes in the metal are absent. Gaps between plates and holes are also more readily filled than by arc welding. The wide heat affected zone and high distortion persists and suitable precautions have to be taken.

The most widely used combination of gases is oxygen and acetylene since only these give a flame temperature and rate of heat input sufficiently high to melt steel. Combustion takes place in two stages (Figure 68). There is an inner reaction flame where the oxygen and acetylene react together to produce carbon monoxide and hydrogen. This reaction is highly exothermic. It can be regarded as a controlled explosion which is attempting to travel through the gas mixture down the torch tubes, but is prevented from doing so by the velocity of the gas stream flowing out of the torch and thus remains stabilized as a flame at the torch tip. Immediately in front of the inner flame there is the region of highest temperature, i.e. approximately 3500 °C. Carbon monoxide and hydrogen are themselves combustible with oxygen, and as they travel away from the inner flame into the surrounding atmosphere a further flame is formed, known as the diffusion flame, where they combine with oxygen diffusing inwards from the atmosphere to form carbon dioxide and water vapour. A flame burning equal volumes of oxygen and acetylene is known as a neutral flame. It is reducing in nature, thereby reducing any iron oxide to iron and taking up the oxygen; consequently there is no need to use a flux when welding steel.

Care and handling of gases

Storage
Cylinders should be stored in a cool dry place away from excessive heat or corrosion. Acetylene cylinders should always be stored upright in a place vented to atmosphere with flame-proof switches and lights. Empty cylinders should be kept separate and clearly marked to distinguish them from full ones. Acetylene cylinders should not be mixed with oxygen cylinders. No smoking or naked lights are allowed in the store, which should be easily accessible with fire extinguishers available (carbon dioxide type).

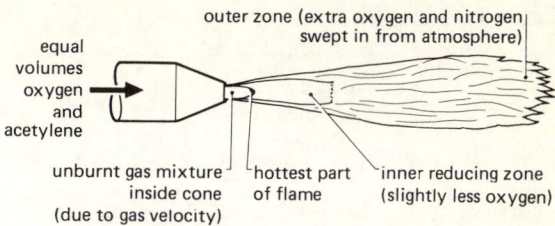

Figure 68 *The oxy-acetylene flame, showing the various zones*

Safe handling
Cylinders should not be dropped from a height or used as rollers, work supports or jacks. Cylinders should be used upright and safely secured away from direct arcing on the cylinder or a naked flame. Do not allow grease or oil to come in contact with cylinders, especially compressed oxygen, as this may cause an explosion. Hydrogen is released from the breakdown of the oil and causes the explosion.

Explosions
Some causes of explosions when using gases are:

1. Leaking cylinders, torches, gauges and connections.
2. Inadequate ventilation in confined places.
3. Transferring gas from one cylinder to another.
4. Dropping and mishandling cylinders and equipment.
5. Leaving lighted torches unattended.
6. Using compressed oxygen as a means of ventilation or compressed air.
7. Allowing hot metal and sparks to fly on to hoses and connections.
8. Testing for leaks with a naked flame.

Flashback may be prevented by fitting preflash flashback arrestors which prevent backflow of gases by means of a non-return sintered stainless steel flame trap. Hose check valves fitted at the blowpipe end have a spring-loaded valve which seats instantly the gas flow is reversed. The principle of using a colour scheme for identifying cylinders holding gases in common use is that

Table 1 *Identification of cylinders*

Gas	Ground colour of cylinder	Typical use
Acetylene	Maroon	Welding, cutting
Air	Grey	Brazing
Argon	Blue	TAGS, MAGS, PlC
Carbon dioxide	White strip / Syphon type / Black	MAGS
Helium	Med. brown	TAGS, MAGS
Hydrogen	Red	Welding, cutting
Methane	Red	Welding, cutting
Nitrogen	Dark grey (black neck)	TAGS
Oxygen	Black	Welding, cutting

TIG or TAGS = tungsten arc gas-shielded welding
MIG or MAGS = metal arc gas-shielded welding
PlC = plasma cutting

yellow should represent toxic or poisonous gases and red or maroon inflammable gases (Table 1). Some gas cylinders have a distinguishing colour band painted around the neck or down the length. Manufacturers often paint an aluminium panel on the cylinder body to show up special markings, and also attach identification labels.

Oxy-acetylene welding equipment

High pressure oxy-acetylene welding systems consist of the following equipment (see Figure 69):

1 A supply of dissolved acetylene (DA) stored in steel cylinders which contain a porous substance (charcoal) and a solvent (acetone) for the gas. The cylinders are charged to a pressure of approximately 15.5 bars (221.4 psi). There are various sizes, the usual ones being 3.39 m^3 and 5.66 m^3.
2 A supply of oxygen gas in alloy steel cylinders charged to a pressure of 172.5 bars (2464.3 psi). Single cylinders of oxygen and DA or a bank of cylinders called a manifold may be used. A manifold supplying several blowpipes usually has an acetylene safety valve and two line pressure gauges. Copper pipe is never used for conveying acetylene because an explosion compound is formed.
3 Pressure regulators for each gas to reduce the cylinder pressure to a suitable value for welding. The regulator pressure screw should always be slackened off after welding has finished.
4 Rubber canvas hose with special connections. Hoses for fuel gas are red with left-handed connections, whilst the oxygen hose is blue with right-handed connections.
5 Blowpipe with set of nozzles. The nozzle size may indicate the approximate consumption of the gas in litres/hour using a neutral flame.

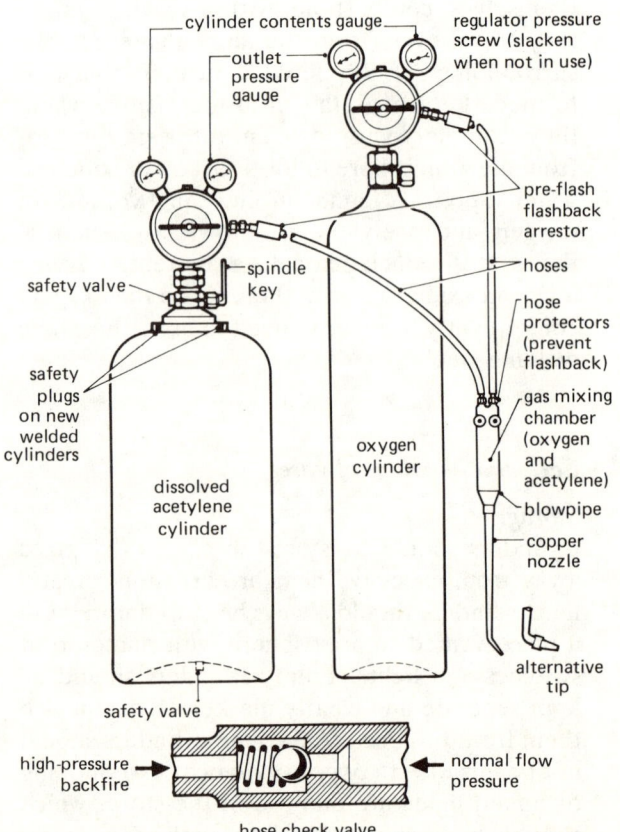

Figure 69 *Oxy-acetylene welding equipment*

Welding processes

6 Special tinted welding goggles.
7 A spark lighter.

Discharge rates of cylinders

The hourly withdrawal rate of gas from an acetylene cylinder must not be greater than 20 per cent (one-fifth) of its contents, otherwise acetone may be drawn off. The acetylene cylinder should always be stood upright. Excessive withdrawal rates of oxygen, especially in cold weather, can cause 'icing up' of regulators. Hot water should be used to thaw out frozen valves and regulators, *not* naked flames.

Cylinders

Oxygen cylinders are painted black and are fitted with a valve outlet which has a right-hand thread. Acetylene cylinders are painted maroon and have a valve outlet with a left-hand thread. They may have a protective cover for the valve.

Regulators

Each cylinder must be fitted with the appropriate regulator which:

1 Shows the cylinder pressure.
2 Shows the pressure at the regulator outlet.
3 Enables accurate control of the output pressure using the handwheel.
4 Prevents gas from flowing back into the cylinder.
5 Prevents the gas in the cylinder from being ignited by a 'flashback'.

The regulator is set for zero output when the regulating screw is turned fully anti-clockwise, i.e. unscrewed.

Gas economizer

This consists of two valves operated by a lever. It is fitted between the regulators and the blowpipe; when the blowpipe is hung on the level, both gas supplies to the blowpipe are cut off. It is normal for a flame to be adjacent for reignition when using an economizer. This enables:

1 Saving of gas, without disturbing the valve adjustments.

2 Greater safety, by ensuring that the flame is extinguished when the blowpipe is not in use.

Blowpipe

This mixes the oxygen and acetylene before they are burned at the nozzle. It is fitted with control valves to adjust the rate of flow of the gas; this varies:

1 The intensity of the flame.
2 The character, i.e. oxidizing and carburizing of the flame, which is determined by the relative proportions of the two gases. Different sized nozzles may be fitted to the same blowpipe. Cutting blowpipes have a lever which when operated supplies the additional oxygen required for cutting.

Nozzles

The size of nozzle controls the size of the flame. For heavier work, a larger nozzle is needed. The gas pressures at the regulators must be set at the correct values for the particular nozzle in use.

Assembly of equipment

Stand both cylinders vertically either in a cylinder trolley or in cylinder stands. Cylinders should never be used lying on the floor. Ensure that jointing surfaces of cylinder valves are free from oil or grease and momentarily open, then close valves to blow out any dirt or obstructions.

Screw the appropriate regulator into each cylinder and tighten. Excessive tightening is not necessary but make certain that joints are gas-tight. The threads on oxygen cylinder valves and regulators are right-handed and on acetylene equipment left-handed.

Blow through the hoses with air to remove any dirt or dust before connecting to the regulator. Oxygen should not be used for this purpose. Connect the hoses (acetylene red, oxygen blue) to the threaded outlet of the regulator by the screwed connectors secured to the hose ends. Check that the control valves on the blowpipe are closed and screw the appropriate size of nozzle into the blowpipe (Table 2).

Table 2 Welding data: high-pressure blowpipes

Low-carbon steel thickness			Nozzle size	Operating pressures				Gas consumptions			
				Acetylene		Oxygen		Acetylene		Oxygen	
mm	in	swg		bar	lbf/in^2	bar	lbf/in^2	l/h	ft^3/h	l/h	ft^3/h
0.9	—	20	1	0.14	2	0.14	2	28	1	28	1
1.2	—	18	2	0.14	2	0.14	2	57	2	57	2
2	—	14	3	0.14	2	0.14	2	86	3	86	3
2.6	—	12	5	0.14	2	0.14	2	140	5	140	5
3.2	⅛	10	7	0.14	2	0.14	2	200	7	200	7
4	5/32	8	10	0.21	3	0.21	3	280	10	280	10
5	3/16	6	13	0.28	4	0.28	4	370	13	370	13
6.5	¼	3	18	0.28	4	0.28	4	520	18	520	18
8.2	5/16	0	25	0.42	6	0.42	6	710	25	710	25
10	⅜	4/0	35	0.63	9	0.63	9	1000	35	1000	35
13	½	7/0	45	0.35	5*	0.35	5*	1300	45	1300	45
19	¾	—	55	0.43	6	0.43	6	1600	55	1600	55
25	1	—	70	0.49	7	0.49	7	2000	70	2000	70
25+	1+	—	90	0.63	9	0.63	9	2500	90	2500	90

Lighting-up procedure

Use a cylinder key and slowly open the valves on both cylinders one complete turn. Do not open the valves suddenly as pressure surges may damage regulators and could cause an accident. Adjust the regulators to the required working pressures. Always ensure adequate supplies of gas for maximum safety. Purge hoses free from mixed gases (i.e. air or oxygen in the fuel gas hose or fuel gas in the oxygen hose) by opening each blowpipe valve separately for a few seconds and then closing again. Open the acetylene control valve on the blowpipe, check the working pressure and wait for a few seconds until air is blown out and pure acetylene is emerging from the blowpipe nozzle, then light the gas with a spark lighter.

Adjust the blowpipe valve until the flame just ceases to smoke then gradually turn on the blowpipe oxygen control valve until the white cone of the flame is sharply defined with the merest trace of acetylene haze. In this condition the flame is neutral and is burning approximately equal volumes of oxygen and acetylene. It is advisable to have a slight haze of acetylene around the centre cone because there is a tendency for the flame to become slightly oxidizing as welding proceeds and in most cases it is harmful to the weld to have an excess of oxygen. If an oxidizing flame is required reduce the flow of acetylene by turning down the acetylene control valve until the correct conditions are obtained. For a carburizing flame, increase the flow of acetylene to produce a feather of acetylene at the end of the centre white cone.

Welding flame conditions

There are three distinct flame settings (see Figure 70):

(a) *Neutral flame* burns equal quantities of oxygen and acetylene. (In practice, it is advisable to have the slightest possible acetylene haze at the cone tip to begin with.)

(b) *Carburizing flame* has an excess of acetylene which results in a carbon-rich zone extending around and beyond the cone.

   Note: Both the neutral and carburizing flames are reducing in nature.

(c) *Oxidizing flame* has an excess of oxygen which results in an oxygen-rich zone just beyond the cone. This flame is obtained by setting to neutral and then turning the fuel gas down.

Applications of the various flame settings
Neutral flame
This flame is used for the welding of steel, cast iron, stainless steel, copper and aluminium.

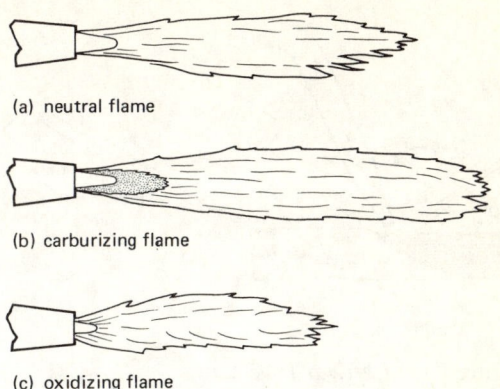

Figure 70 *Welding flame conditions*

Carburizing flame
This flame is used for hard surfacing rod. The parent metal is heated and the flame gives up carbon to the parent metal so lowering its melting point and allowing the rod to be deposited quickly without deep penetration. A very slightly carburizing flame is often used for non-ferrous metals where the smallest amount of oxygen would be undesirable.

Oxidizing flame
This is undesirable when strongly oxidizing, except for welding brass. When slightly oxidizing it is used when brazing zinc-coated sheet.

Shutting-down procedure

Shut off acetylene first by closing the blowpipe control valve then follow with closure of the oxygen valve. Close the supply valves on cylinders. Then open and close the blowpipe valves one at a time to relieve pressure in the hoses – oxygen first, then acetylene. Wind back the pressure adjusting screws on both oxygen and acetylene regulators.

Welding techniques
Leftward welding
Leftward welding (Figure 71) is used on steel for flanged edge welds, for unbevelled plates up to

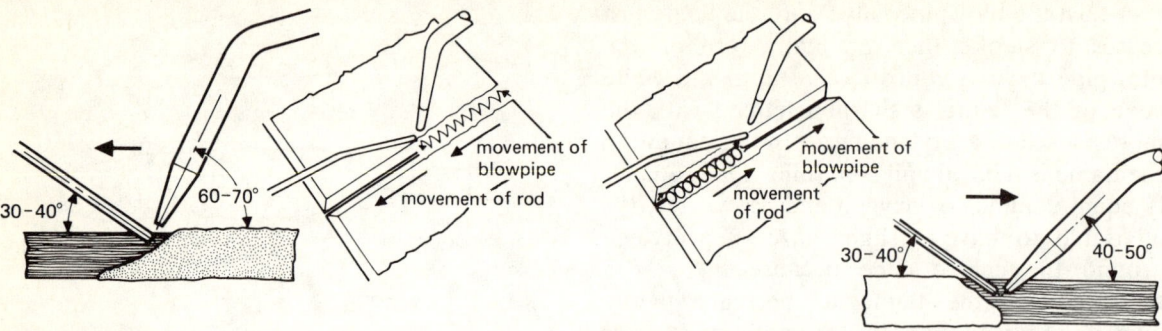

Figure 71 *Leftward welding*   Figure 72 *Rightward welding*

3.2 mm (⅛ in) and for bevelled plates up to 4.5 mm (³⁄₁₆ in). It is also the method usually adopted for cast iron and non-ferrous metals. Welding is started at the right-hand end of the joint and proceeds towards the left. The blowpipe is given a forward motion with a slight sideways movement to maintain melting of the edges of both plates at the desired rate and the welding rod is moved progressively along the weld seam. The sideways motion of the blowpipe should be restricted to a minimum.

Rightward welding

Rightward welding (Figure 72) is recommended for steel plate over 5.0 mm (³⁄₁₆ in) thick. Plates from 5.0 mm to 8.0 mm (³⁄₁₆ to ⁵⁄₁₆ in) need not be bevelled; over 8.0 mm (⁵⁄₁₆ in) the edges are bevelled to 30 degrees to give an included angle of 60 degrees for the welding V. The technique is suitable for horizontal/vertical position.

The weld is started at the left-hand end and moves towards the right with the blowpipe flame preceding the filler rod in the direction of travel. The rod is given a circular forward motion and the blowpipe is moved steadily along the weld seam. This is faster than leftward welding; it consumes less gas, the V angle is smaller, less filler rod is used and there is less distortion.

The all position rightward technique is a modification of the above and is particularly suitable for low-carbon steel plate and pipe in the vertical and overhead positions. The advantages are that it enables the welder to obtain a uniform penetration bead and an even build-up, particularly in fixed position welding; the welder can work with complete freedom of movement and has a clear view of the weld pool and the fusion zone of the joint. Considerable practice is required to become familiar with this technique even by operators skilled in normal downhand rightward welding. An apparent undercutting of the plate surface at the edges of the weld bead is a fault to which this technique is prone but this can be controlled by appropriate manipulation of the rod and flame. The rod and blowpipe angle should be adjusted to give adequate control of the molten metal as in normal rightward welding.

Vertical welding

Vertical welding (Figure 73) may be used on unbevelled steel plate up to 3 mm (⅛ in) thickness and up to 15 mm (⅝ in) when two welders are employed working on both sides of the joint; welding starts at the bottom and proceeds vertically.

Edge preparations

Edge preparations are detailed in Figure 74.

Welding various metals

Low-carbon steel

Select the appropriate method and always use a neutral flame. Fluxes are not necessary. An

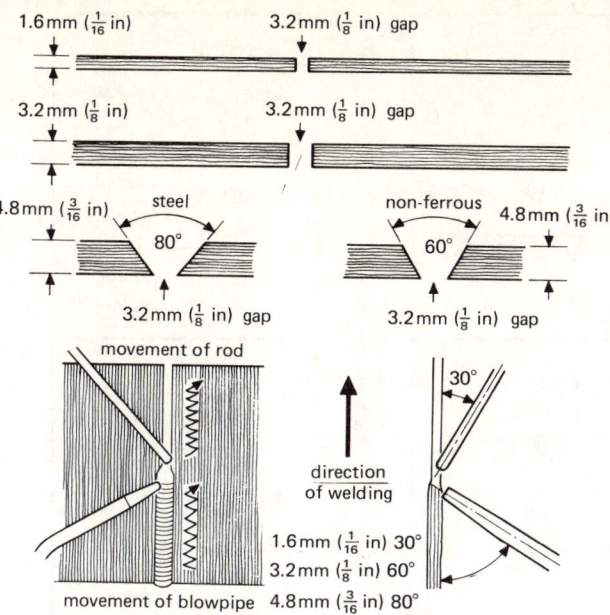

Figure 73 *Vertical welding, single operator, for plate thicknesses up to 4.5 mm ($^3/_{16}$ in). No movement of blowpipe on plate up to 3.2 mm ($^1/_8$ in) thickness; slight movement of blowpipe as above on plate thicker than 3.2 mm ($^1/_8$ in) thick*

improvement in the physical properties of welds in some steels is obtained by normalizing the weld immediately upon completion.

Stainless steel
A welding rod having a composition suitable for welding stainless steel should be used. A flux can be applied to the underneath surface of the metal along the seam and a neutral flame should be used. Keep the welding rod in the flame the whole time and upon completion withdraw the flame slowly from the weld to avoid cracking. Do not interrupt the weld and carry out the work as quickly as possible. Carefully remove all oxide and scale when the job is finished. The weld may be ground and polished.

Cast iron
Use a welding rod high in silicon to obtain a deposit of grey cast iron. Use cast iron flux and preheat the metal to a dull red before starting. Take precautions to allow for the effects of expansion and contraction and do not hurry either the preheating or subsequent cooling; a preheating furnace will be necessary for all but the smallest articles. Make certain that the metal in the neighbourhood of the weld is clean and free from rust, scale or grease before welding commences.

Malleable iron
Do not attempt to use a cast iron rod for welding malleable iron. The best process is to bronze weld using a nickel bronze rod and flux.

Aluminium
Carefully remove all grease or oil and clean the edges with a wire brush. For pure aluminium sheet, either paint both sides of the metal with flux or dip the hot rod into the flux and allow it to coat the rod like a varnish. Use a soft neutral flame and tack at frequent intervals before commencing to weld. Welding must be carried out quickly, and if the material thickness warrants it the vertical method should be used. For aluminium castings use a cast aluminium alloy rod with flux and preheat before welding; do not attempt to weld before the casting reaches a suitable temperature. When the temperature of the casting is sufficient to char wood or sawdust it is correct for welding. Melt the rod well into the weld and do not rely upon the heat of the molten metal to fuse the aluminium. Use the welding rod to puddle the molten metal and cool very slowly. For aluminium tubes use pure drawn aluminium rod or 5 per cent silicon alloy but do not use this alloy in corrosive conditions. Prepare in the normal manner and wash thoroughly when welding is complete to remove all flux residue.

Aluminium brazing
The filler rod must have a lower melting point than the parts to be joined and a 10–13 per cent silicon aluminium alloy rod is used; this gives a temperature latitude of 73 °C between its melting point and that of pure aluminium. Press

Thickness of metal	Diameter of welding rod	Edge preparation		Speed mm/min	Thickness of metal
Less than 0.9 mm (20 swg)	1.2–1.6 mm ($\frac{3}{64}$–$\frac{1}{16}$ in)		leftward welding	127–152	0.8 mm ($\frac{1}{32}$ in)
				100–127	1.6 mm ($\frac{1}{16}$ in)
0.9–3 mm (20 swg–$\frac{1}{8}$ in)	1.6–3 mm ($\frac{1}{16}$–$\frac{1}{8}$ in)	0.8–3 mm ($\frac{1}{32}$–$\frac{1}{8}$ in)		100–127	2.4 mm ($\frac{3}{32}$ in)
				90–100	3 mm ($\frac{1}{8}$ in)
3–5 mm ($\frac{1}{8}$–$\frac{3}{16}$ in)	3–3.8 mm ($\frac{1}{8}$–$\frac{5}{32}$ in)	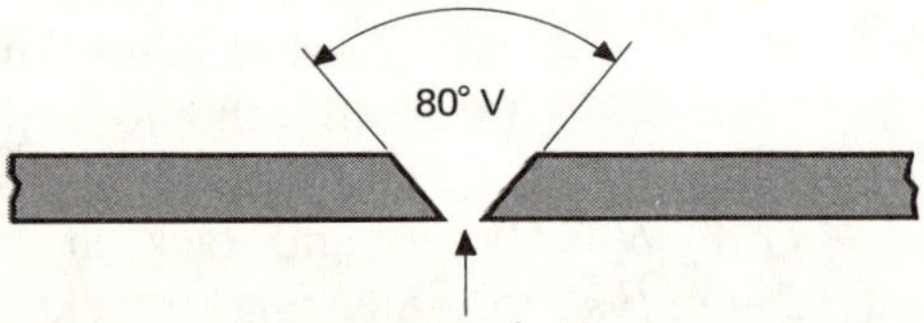 80° V; 1.6–3 mm ($\frac{1}{16}$–$\frac{1}{8}$ in)		75–90	4 mm ($\frac{5}{32}$ in)
				60–75	4.8 mm ($\frac{3}{16}$ in)
5–8.2 mm ($\frac{3}{16}$–$\frac{5}{16}$ in)	3–3.8 mm ($\frac{1}{8}$–$\frac{5}{32}$ in)	3–3.8 mm ($\frac{1}{8}$–$\frac{5}{32}$ in)		50–60	6.4 mm ($\frac{1}{4}$ in)
			rightward welding	35–40	8 mm ($\frac{5}{16}$ in)
8.2–15 mm ($\frac{5}{16}$–$\frac{5}{8}$ in)	3.8–6.5 mm ($\frac{5}{32}$–$\frac{1}{4}$ in)	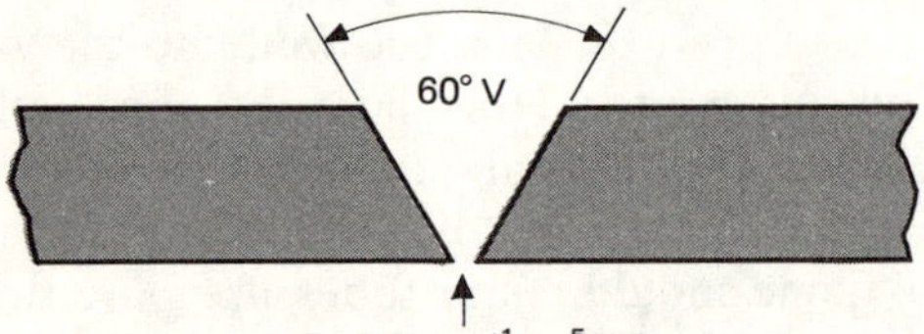 60° V; 3–3.8 mm ($\frac{1}{8}$–$\frac{5}{32}$ in)		30–35	9.5 mm ($\frac{3}{8}$ in)
				22–25	12.5 mm ($\frac{1}{2}$ in)
15 mm ($\frac{5}{8}$ in) and over	6.5 mm ($\frac{1}{4}$ in)	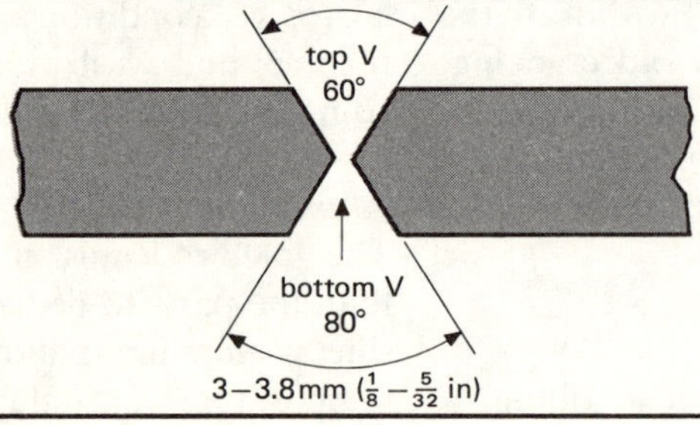 top V 60°; bottom V 80°; 3–3.8 mm ($\frac{1}{8}$–$\frac{5}{32}$ in)		19–22	15 mm ($\frac{5}{8}$ in)
				15–16	19 mm ($\frac{3}{4}$ in)
				10–12	25 mm (1 in)

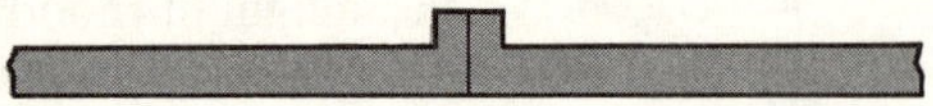

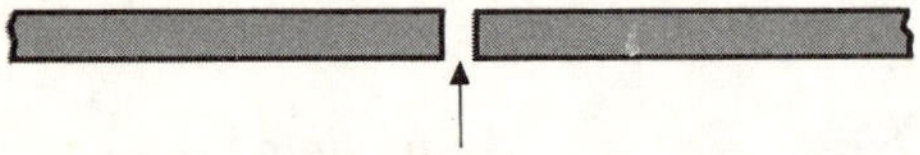

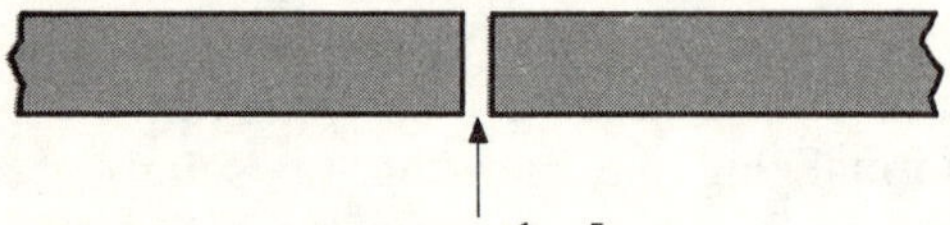

Figure 74 *Edge preparations*

fits and close tolerances are not recommended; slight clearances are advisable to allow complete penetration of the brazing metal and flux. Adequate support is necessary for long seams to prevent sagging. Commercially pure aluminium sheet can be brazed without mechanical removal of the normal surface shine or oxide film providing it is clean and free from other contamination. Aluminium alloys suitable for brazing must have the surface oxide film removed mechanically by brisk scratching with steel wool, wire brush or file. The correct aluminium brazing flux should always be used. The end of the rod is heated and dipped into the flux and the 'tuft' of the flux adhering to the rod is then touched down upon the surface of the joint to check the temperature. At the correct temperature the flux will begin to flow smoothly and rapidly forward along the joint. It is most important that no filler rod is melted until the flux can be seen to be flowing freely induced by the heat of the work. Capillary attraction with the driving action of the flame moving close to the joint surface causes the molten metal to run forward along the joint. Remove all flux deposits with warm water as quickly as possible after brazing. If possible follow this by a dip in a weak acid solution, then rewash with warm water.

Copper

Use a welding nozzle two sizes larger than for the same thickness of steel and align the edges to be welded with lightly adjusted clamps or jigs; do not tack weld or restrain movement entirely. A special deoxidized copper welding rod is necessary, if possible containing some constituent to improve metal fluidity. Flux is not essential. Make certain by preheating that the metal in the neighbourhood of the weld is hot enough to give adequate fusion of the edges and not just adhesion of the deposited metal. Vertical welding gives excellent results on copper and should be used wherever possible. Hammering the weld whilst it is still hot and before the temperature has fallen below 600 °C improves mechanical properties, consolidates the surface and removes porosity. Do not hammer copper when 'black hot'.

Copper tube can be fusion welded but bronze welding is more usual. When bronze welding light-gauge copper tubes, the end of one tube is prepared by opening it out to form a bell-mouth into which the other tube is inserted. Use silicon bronze welding rod with flux. Preheat the joint area with a neutral flame and touch the flux end of the filler rod into the joint; when it is seen to 'wet' the surface, adjust the flame to a suitable oxidizing condition and deposit the rod with a regular welding deposition technique. Avoid overheating the joint and tube wall.

Brass and bronze

An oxidizing flame is necessary for welding most brasses, especially tin-bearing bronze. If a neutral flame is used on copper–zinc alloys, porosity and poor mechanical properties may result. To adjust the flame take a small sample of the brass to be welded and melt it with a neutral flame – some fuming will occur. Cut down the acetylene flow so that there is an excess of oxygen in the flame and fuse a further sample of brass; continue decreasing the acetylene flow until a sound deposit is obtained. When the flame is suitably adjusted it will be seen that the inner cone is slightly shorter and more pointed than the normal neutral cone. Prepare the edges of the brass or bronze in the normal manner and weld, using a bronze welding rod of a type recommended for the metal being welded. Make certain that the edges of the seam are at the point of melting before welding begins.

Bronze welding

Bronze welding involves the use of alloy bronze rods and is used for making joints in copper, for joining dissimilar metals, and for repairs to cast iron. Among the rods used are silicon bronze, nickel bronze and manganese bronze with flux.

It is essential that the edges of the materials should not be melted but merely raised to red heat. For example, when bronze welding cast iron a manganese bronze rod should be used; this melts at 890 °C, well below the melting point of cast iron. The fracture, or the edges of the metal to be joined, are prepared in the usual manner, care being taken to ensure that all sharp edges are rounded off and that they are absolutely clean. The joint so formed has excellent mechanical properties and providing there is no objection to dissimilarity in colour, bronze welding can be recommended for a variety of purposes. Fractured castings should be preheated before welding.

Depositing hardfacing rods

This process (Figure 75) consists of laying down a hard deposit which is very resilient to wear on a steel or cast iron surface. For steel, the flame is adjusted to an excess of acetylene to give a feather 1.5 to 2.5 times the length of the normal neutral inner cone. The steel is raised to a dull red heat and when it begins to 'sweat' the rod is deposited on the surface. Upon completion the work must be cooled slowly either in a furnace or in lime or mica dust. The deposition of hardfacing on cast iron is not as successful as on steel. Since cast iron cannot be made to 'sweat' it is advisable first to fuse a layer of hardfacing rod on to the surface as in the case of welding using cast iron flux, then to 'sweat' a second layer on to the previous deposit. Rods containing tungsten carbide are also deposited with this technique.

Building up worn surfaces

For building up surfaces subject to a large amount of wear it is usual to deposit a rod which has special wear-resisting properties. An excess acetylene flame is used; the base metal is preheated until it begins to 'sweat' and then the special wear-resisting alloy steel rod is melted on to the surface. The weld metal should be added in small deposits and the whole surface gradually built up to the required thickness. To obtain the maximum wear-resisting qualities, the weld metal should be thoroughly hammered after completion of the weld. In addition to the special wear-resisting alloy steel there is a wide range of high-carbon and alloy steel rods which, with appropriate flame adjustment, may be used for building-up purposes.

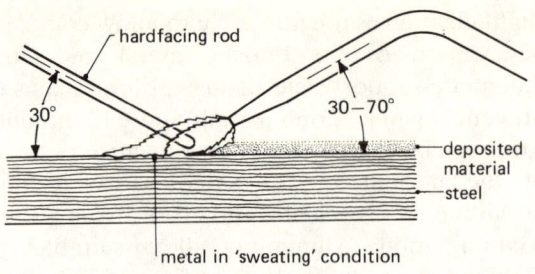

Figure 75 *Hardfacing procedure*

Oxygen cutting

This process utilizes the fact that heated iron oxidizes easily and that if heated to a temperature of 870 °C in the presence of an ample supply of oxygen it will burn like any other combustible material. Should a length of iron or steel be heated in a forge, it will react with the oxygen in the atmosphere when withdrawn from the fire. If this process were to be continued over a period of time, it would be found that the metal would be completely consumed, being converted into iron oxide. This is a process of oxidation and oxygen cutting is a chemical action. This process is usefully applied in oxy/fuel-gas cutting where the metal is quickly brought to its combustion temperature by the use of a large-capacity preheating nozzle, and then a jet of pure oxygen is directed on to the heated area. This has the effect of producing rapid oxidation, and the oxide is then melted and blown away by the force of the high-velocity oxygen jet. The hole produced will correspond in size to the diameter of the orifice in the nozzle. If the nozzle is moved forward at a regular speed, a section of the plate will be removed; this cut is called a 'kerf'.

A metal cannot be cut by the oxy/fuel-gas process unless two essential conditions can be satisfied:

1. Oxidation of the metal must take place at a temperature below the melting point of the metal.
2. The oxide produced must have a melting point below that of the metal.

Many metals cannot satisfy these two conditions, namely non-ferrous metals, stainless steels and cast irons. Almost all other steels do meet these criteria and can be cut by the oxy-gas process.

Heating prior to cutting (preheating) is always carried out with a flame adjusted to neutral. There are several types of nozzle but all are designed on the same principle, i.e. a central orifice which carries the high-purity high-pressure cutting oxygen and surrounded by one or more gas ports carrying a mixture of fuel gas and oxygen.

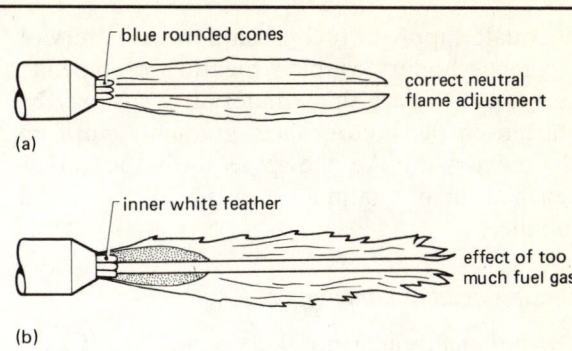

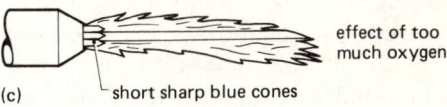

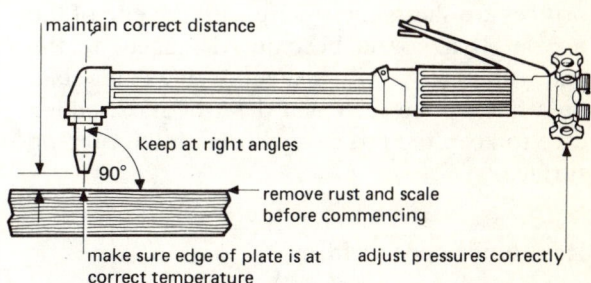

Figure 76 *Flame adjustment*

Figure 77 *Oxygen cutting*

Lighting-up procedure

Use a cylinder key and slowly open the cylinder valves one complete turn. Do not open the valves suddenly as pressure surges could cause an accident. Adjust regulators to the required working pressures. Purge hoses free from mixed gases (i.e. air or oxygen in the fuel gas or fuel gas in the oxygen hose) by operating each blowpipe valve separately for a few seconds, then closing them again. Open the fuel gas control valve on the blowpipe, and light the gas with a spark lighter. Open the oxygen valve and adjust the heating flame to the conditions required (Figure 76). Depress the cutting oxygen lever and readjust the heating flame to the required condition.

Cutting technique

Hold the blowpipe with the nozzle at right angles to the plate and apply the heating flame to the edge of the material furthest from the operator (Figure 77). A spade or roller guide attached to the nozzle is of assistance in holding the blowpipe steady. When the edge of the metal attains a bright red heat, operate the cutting oxygen lever and draw the blowpipe towards the body along the line of cut. The nozzle is moved progressively at a constant height (called stand-off distance) and speed in the direction of cutting. The nozzle should always be at 90 degrees to the surface unless bevel cutting is being carried out. The tips of the blue cones should be kept at about 3 to 5 mm above the plate surface.

Thick material

Because of the heavy gas consumption when cutting thick material, ensure that there is an adequate supply of fuel gas and use a battery of oxygen cylinders coupled together or pipeline supplies. On very thick material, a cut may be started on the bottom face; gradually work up the edge with the blowpipe until the top is reached then continue the cut in the usual manner.

Cutting nozzle maintenance

Do not maltreat a nozzle; do not use it as a hammer or lever. To clean nozzle orifices, sets of special nozzle cleaning reamers are available. After prolonged use the nozzle may become dirty and it should be immersed in nozzle cleaning compound according to the directions supplied. Effective preheat shape and cutting oxygen stream can only be maintained if gas orifices are sharp and square with the end of the nozzle. If a nozzle becomes damaged on the end, rub it down with a sheet of fine emery laid on a flat surface such as a sheet of glass, taking care to keep the nozzle square with the rubbing surface.

Resistance spot welding

The mass production of motor car bodies involves the use of spot welding and motor cars are held together by between 4000 and 9000 spot welds. Spot welding is a variant of electric resistance welding and this method of assembly is used not only because of its technical advantages but also because of the important cost reduction that can be achieved. The advantages of spot welding are the following:

1. Spot welding is the fastest, strongest and most reliable method of joining two pieces of sheet metal.
2. Spot welding gives little or no metal distortion owing to too great radiation of heat; the heat required for making a spot weld is limited to a small area around it.
3. In some cases it is possible to make spot welds which are almost invisible. This is specially valuable when welding eye-catching parts like radiator grilles.
4. No preparatory work is needed for spot welding as compared to bolting or riveting. There is no need to drill metal, screw bolts or drive rivets, and no need for electrode material, oxygen or acetylene; hence there is a considerable cost reduction.
5. Spot welding does not alter the mechanical characteristics of the parent metal. Good spot welds are equal in strength, hardness and ductility to the original specification of the sheet metal as laid down by the designer for its mechanical performance.

Although repair work is different from production, the repairer should be able to carry out work to standards of quality identical to those obtained in mass production and make use of the advantages offered by spot welding.

The principles of spot welding

A comparison can be made between an electrical spot weld (Figure 78) and a forge weld for in both of these processes a union is formed by an amalgamation of the molecules. In the case of forge welding, the metal pieces are heated in the forge furnace and then hammered until homogeneous. In the case of spot welding the pieces are heated electrically using the same principle as in an electric radiator (resistance effect); they are subjected to a certain pressure which must be sufficient and continuous to enable the molecules i.e. submicroscopically small particles, to interlock.

Spot welding is a self-contained weld method on a limited surface area without added weld metal. It involves:

1. The use of a preset pressure.
2. An electric current of precise intensity and duration.

A spot welding gun (Figure 79) must have:

1. A pressure device worked by the operator for transmitting the pressure to the electrodes.
2. A transformer to enable current at high intensity to be fed to the electrodes.

Welding processes

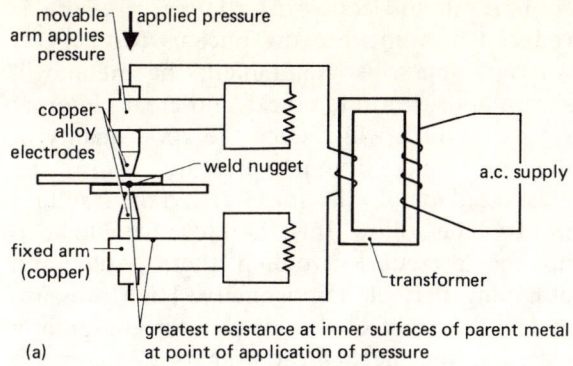

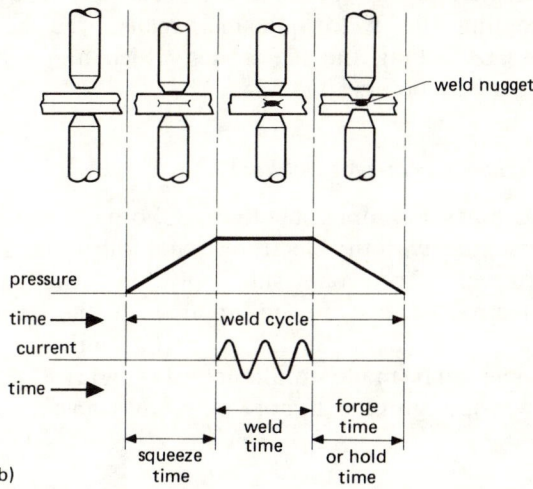

Figure 78 *(a) Resistance spot welding system and (b) relationship between weld formation, current and pressure in welding*

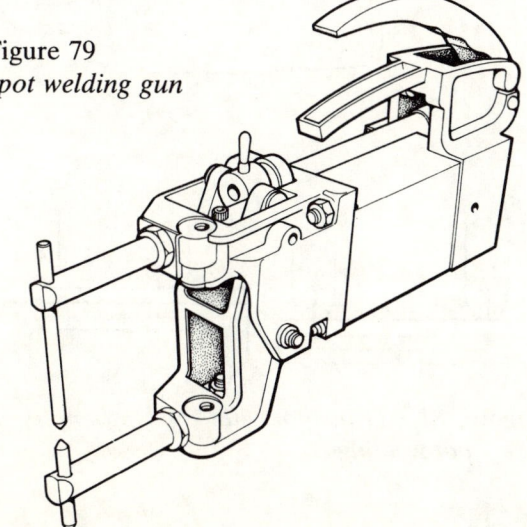

Figure 79
Spot welding gun

These electrodes are made of a copper chrome alloy, selected for strength and good conductivity. The electrodes must have the ability to conduct mechanical and electrical power. The following are the three stages in making a spot weld (see Figure 78):

1 Squeeze time
2 Weld time (duration of current flow)
3 Hold time

Squeeze time
Bringing the parts together is the stage before welding. It is using the electrodes to put strong pressure on the parts to be welded which means, from the mechanical point of view, making contact between the parts to be welded, especially an intimate contact at the exact point where the spot weld will be formed. From the electrical point of view there is better current flow, as the pressure lowers the resistance created by the surfaces in contact.

Three contact surfaces are formed – two between the electrodes and the sheet metal, and one between the sheets of metal themselves. There is better contact between the electrodes

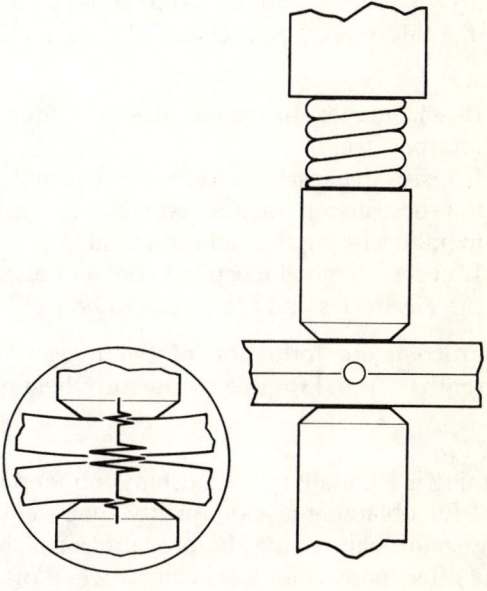

Figure 80 *Contact surfaces*

and the workpieces (copper–steel contact) than between the metal sheets (steel–steel) contact (Figure 80).

The contact between the steel sheets will offer the greater resistance to the flow of electricity. This is indicated by the 'zigzag' in Figure 80. After bringing the workpieces under pressure an electrical connection is made in which the highest resistance appears at the point of contact between the two pieces. Damage to the workpieces and gun will result if the weld current is switched on before full pressure has been reached. When full pressure is reached the second stage, weld time, begins.

Weld time
When the current passes through the assembly it meets with resistance at the points of contact. This resistance will cause heating by opposing the passage of the current. The higher the resistance, the greater the amount of heat generated. The highest resistance is at the point where the two metal sheets are in contact, and it is at this point that the greatest heat is produced. Heat is radiated and plasticizes the metal. This produces a round, flattened nugget surround by a sheath of metal in a plastic condition, forming a sort of crucible holding the hottest metal. During this operation, pressure plays a multiple role:

1. It encourages the molecules of metal to interpenetrate.
2. It resists expansion, and thus resists likewise the destruction of the crucible of plastic metal enclosing the red-hot metal.
3. It ensures a good electrical contact between the electrodes and the pieces to be welded.

After complete formation of the nugget, the current is cut and forging by pressure begins.

Hold time
Forging is a metallurgical finishing job, and it is vital for obtaining a good-quality spot weld. It consists of maintaining the pressure for a short time after the current has been switched off.

When the current is cut, the nugget begins to cool. If, in the course of cooling, pressure is reduced or removed, the nucleus (or nugget) will no longer be maintained; the metal will shrink and form fine cracks and create internal stresses, which will reduce the good quality of the spot weld. If much pressure is lost, the plasticized metal will squirt out and the resultant weld will be hollow. It is, therefore, vital to keep up the pressure, avoiding these faults and obtaining perfect homogeneity. High pressure makes the grain of the metal finer and greatly increases the mechanical strength of the weld. Maintaining the pressure also has the effect of cooling the weld quicker because the heat escapes along the electrodes which are good heat conductors.

Series or twin-spot welding

At times it is impossible to reach both faces of an assembly with the electrodes, and it is necessary to weld from one side only, i.e. the two electrodes have to be applied to the same surface or sheet. By the twin-spot method two spots can be made simultaneously (Figure 81). A twin-spot gun has a more powerful transformer

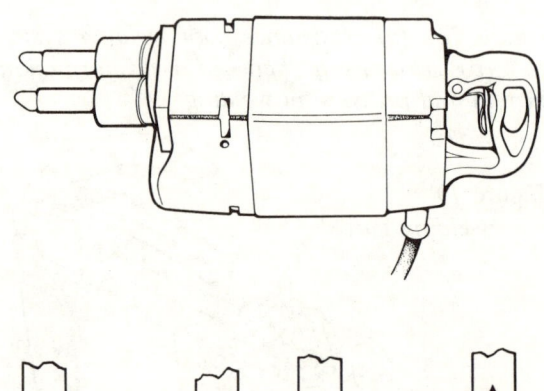

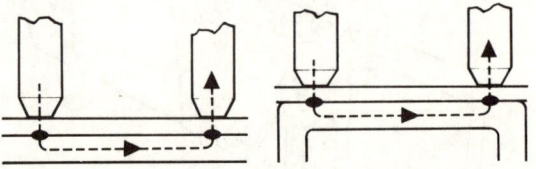

Figure 81 *Twin-spot gun and examples of twin-spot welding*

than a pincer-type gun. It should be fitted with a pressure compensating mechanism to ensure that even weld pressure is exerted by the weld current-carrying electrodes on the workpiece, and it is essential that both electrodes be correctly aligned.

Spot welding quality

The repairer with his spot welding guns must produce first-class quality welds. There is one disadvantage with welding, namely that a visual inspection of a weld gives no guide whatsoever to weld quality. It is therefore vital that the spot welding equipment used in body repair is absolutely reliable and automatically takes care of outside circumstances such as rust, scale, drops in electric mains supply voltage, and general lack of specific experience.

The quality of a spot weld is determined by three factors:

1. The weld must be at least equal to, but preferably should be superior to, the parent metal in tensile strength (and therefore in the homogeneity of the weld nugget).
2. The welding heat must not in any way alter the inherent qualities of the parent metal in terms of strength, hardness and ductility (i.e. flexibility or elasticity).
3. The welding speed must prevent parent metal expansion and contraction so as not to cause panel distortion or create stresses in the assembly.

Resistance welding is different from gas and arc welding in as much as the weld is made at a very much lower temperature and the weld is forged under pressure. A slight shrinkage of the parent metal at the welded spot is natural because a solid weld has been forged between two sheets. This reduces the added thickness of the parent metal without spreading sideways. Equipment for spot welding must therefore produce sufficient current to make a weld quickly; for example, on a 20 + 20 g (1 mm + 1 mm) universally recognized specifications stipulate that not less than 6000 A are available at the welding tips and that the weld time must be not more than $2/5$ s and not less than $1/10$ s. The weld then cools under heavy forging pressure. Resistance spot welding can weld dissimilar metal thickness combinations by using a larger electrode contact tip area against the thicker sheet. This can be done for low-carbon steel having a dissimilar thickness ratio of 3:1. Other resistance welding processes used mainly in production include projection, seam, flash and resistance butt welding.

Metal arc gas-shielded (MAGS) welding

Definition

Gas metal arc welding (Figure 82) is an electric arc welding process which produces coalescence (joining) of metals by heating them with an arc established between a continuous filler metal (consumable) electrode and the work. Shielding of the arc and molten weld pool is obtained entirely from an externally supplied gas or gas mixture.

Metal arc gas-shielded welding may be divided into two types:

1. Metal inert gas (MIG) – used for aluminium, copper, stainless steel and carbon steel. The gas used is an inert one, such as argon or helium.
2. Metal active gas (MAG) – used mainly for carbon steel. The gas used is an active one, such as CO_2, or a gas mixture (it is cheap).

Arc and power polarity

The majority of MAGS applications requires the use of direct current electrode positive (reverse polarity). This type of electrical connection yields a stable arc, smooth metal transfer, relatively low spatter loss and good weld bead characteristics for the entire range of welding current used. Direct current electrode negative (straight polarity) is seldom used, since the arc can become very unstable and erratic even though the electrode melting rate is higher than that achieved with electrode positive. When

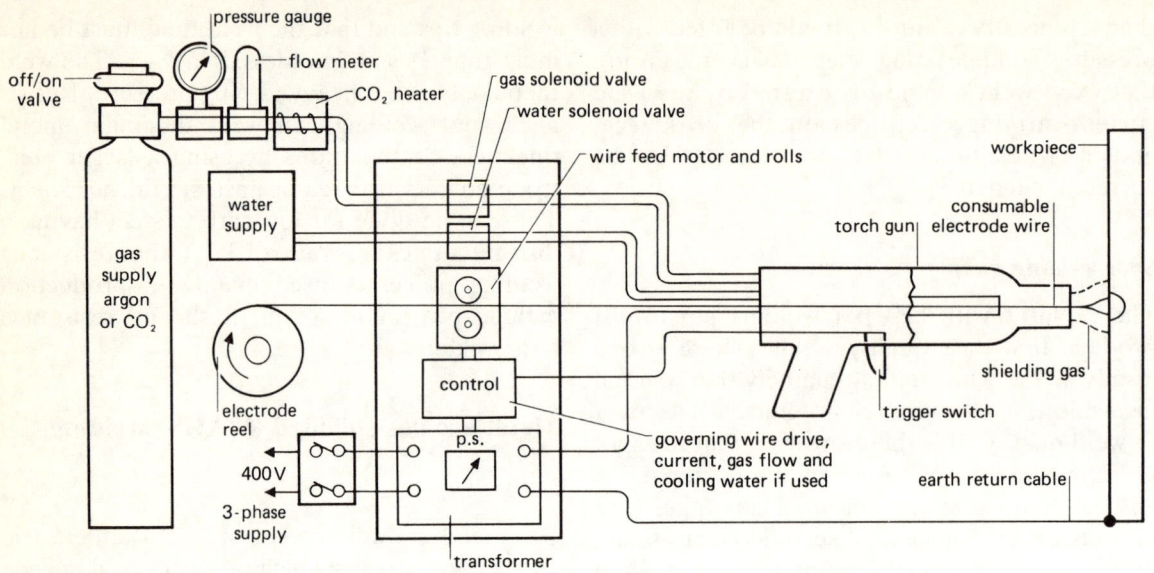

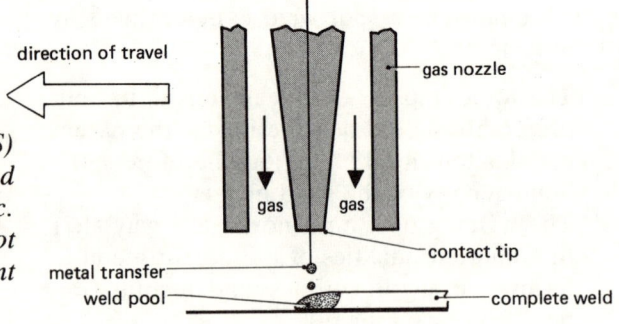

Figure 82 *Metal arc gas-shielded (MAGS) welding. Type of operation: semi-automatic and automatic. Heat source: electric arc, usually d.c. electrode positive. Shielding: gas, which must not be active with the metal being welded. Current range: 60–500 A*

employed, electrode negative is used in conjunction with a 'buried' arc or short-circuiting metal transfer. Penetration is lower with electrode negative than with electrode positive direct current.

Alternating current has found no commercial acceptance with the MAGS process for two reasons:

1 The arc is extinguished during each half-cycle as the current reduces to zero and it may not reignite if the cathode cools sufficiently.
2 Rectification of the electrode positive polarity cycle promotes the erratic arc operation.

Metal transfer

Filler metal can be transferred from the electrode to the work in two ways (see Figure 83):

1 When the electrode contacts the molten weld pool, thereby establishing a short circuit known as short-circuit or dip transfer.
2 When discrete drops are moved across the arc gap under the influence of gravity or electromagnetic forces. Drop (free flight or spray transfer) can be of the following form: projected, repelled or globular. Shape, size, direction of drops (axial or non-axial) and type of transfer are determined by a number of factors such as the following:

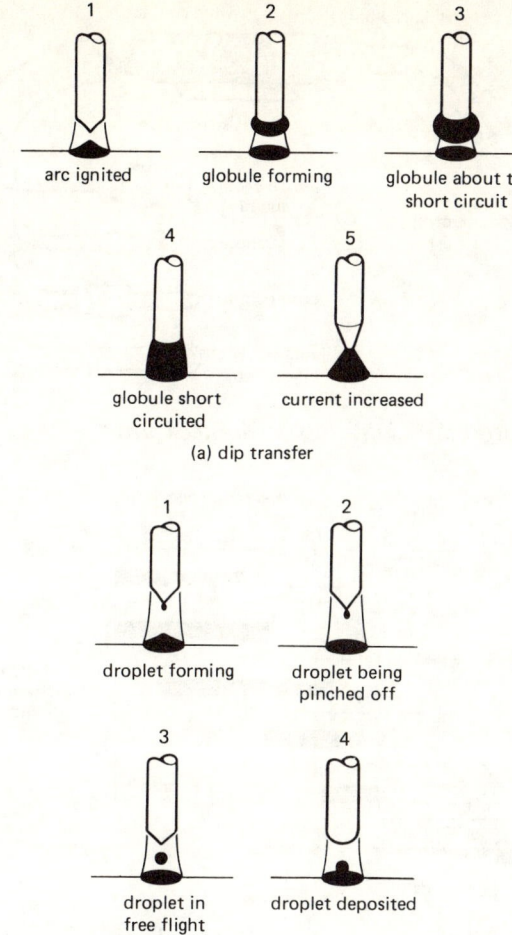

Figure 83 *Metal transfer*

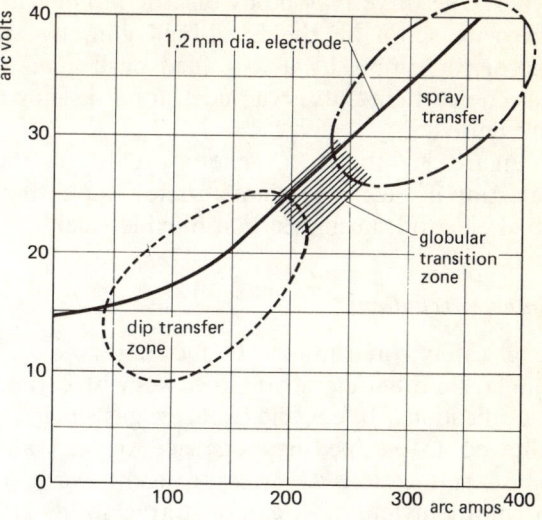

Figure 84 *Graph showing spray and dip transfer ranges with the threshold range in between the two*

(a) Magnitude and type of welding current.
(b) Current density.
(c) Electrode extension.
(d) Electrode composition.
(e) Shielding gas.
(f) Power supply characteristics.

Dip transfer
This is where relatively low arc voltages and currents are utilized to obtain the smooth detachment of droplets from the electrode tip; the electrode diameter and arc voltage are chosen for a selected wire speed to give correct welding conditions. The size of the droplets increases as the current is reduced. 'Stubbing', or welding of the tip to the plate, occurs when current rise is too low. This mode of transfer is used for welding thin plate or sections and for positional work. The sequence of metal transfer is shown in Figure 83(a).

Spray transfer
This mode of transfer is when the tip of the electrode is deposited in the form of a fine spray of molten metal droplets detaching smoothly at a high frequency (Figure 83(b)). True spray transfer occurs at relatively high arc voltages and low currents, but, owing to the low melting point of aluminium, spray transfer can be obtained at relatively low current levels compared with steel. Figure 84 shows a typical graph indicating spray and dip transfer ranges with the threshold range in between the two.

Types of MAGS torches
There are several types of torch, but they may be divided into gas-cooled and water-cooled

types. The drive may be by electric motor with the wire spool on the hand-held gun, by air motor, or simply by a wire-feed push gun. A gas-cooled light-duty swan-neck torch is shown in Figure 85.

Figure 86 shows a general guide to the selection of electrode diameters for low-carbon steel electrodes using carbon dioxide shields.

Safety precautions

The safety precautions to be observed are similar for other metal arc processes with certain modifications. In confined spaces, gas shields if allowed to escape may displace oxygen and cause suffocation. Degreasing agents such as trichloroethylene and carbon tetrachloride decompose around the arc to form poisonous compounds. Local fume extraction should be used when employing very high current densities or flux core electrode wire, and filter breathing pads to prevent inhaling oxide dust. Correct grades of screen glass should be used as ultraviolet light is greater when welding aluminium with an argon shield compared with other processes. Remember to chalk HOT materials after welding, especially aluminium. Adequate protective clothing should always be worn.

Advantages

One of the foremost advantages of the process is versatility. Given the availability by the user of suitable consumable wires (and in some cases, suitable alternative gases) it is possible to change from material to material with minimum delay and using the same equipment. Versatility requires more than this, however, and is often more a matter of type of joint than parent material. In such cases a few adjustments to the controls are generally all that is required. The root, however, of process advantage necessarily lies in the welding economics and the MAGS process provides substantial savings over many forms of welding owing to direct welding speed and indirect time saving. Several other factors which tend to be forgotten favour the MAGS

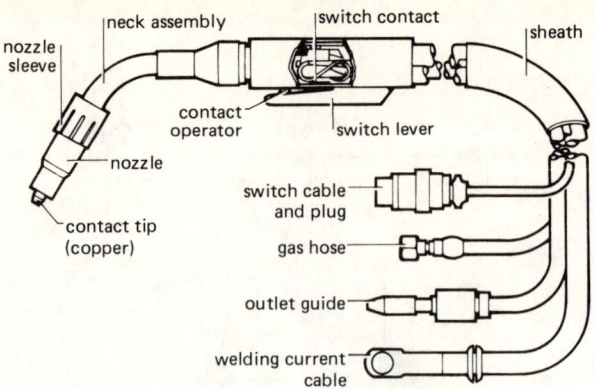

Figure 85 *Light-duty swan-neck torch*

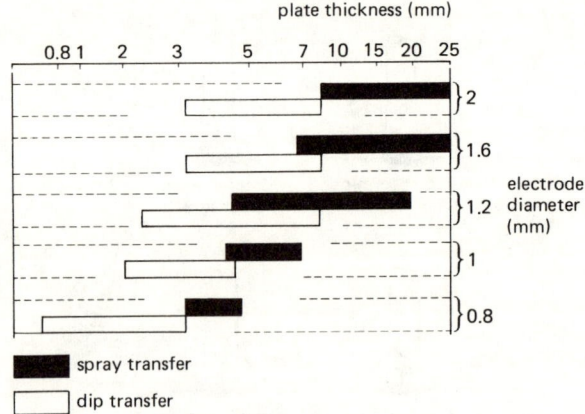

Figure 86 *General guide to selection of electrode diameters for low-carbon steel electrodes using carbon dioxide shields*

welding process. The process is fluxless and therefore produces no welding slag; time spent in removing slag is eliminated. Manual stick electrode welding requires continual replacement of the electrode during which time no welding is taking place. MAGS welding uses a continuous electrode enabling long welding runs to be carried out and the equipment offers a light welding tool which reduces the fatigue experienced by operators. All these indirect savings help to increase the duty cycle of the welding operation. In addition to these attractions there are areas in which MAGS welding has advantages as much associated with process capability as with welding costs. For example, the welding of aluminium has always been a problem; the high oxidization rate makes the use of corrosive fluxes essential with conventional arc and gas techniques. MAGS welding has a self-cleaning characteristic which eliminates the use of fluxes completely.

Metal arc gas-shielded (MAGS) spot welding

An advantage of carbon dioxide MAGS welding is the ability of the process to be adopted for single-side spot welding applications either semi or fully automatically. By extending the welding gun nozzle to contact the workpiece, one-sided spot welds may be performed using dip transfer conditions. Predetermined weld duration times may be employed, the gun trigger being coupled to a suitable timer and, if desired, fully mechanized.

Unlike resistance spot welding, no pressure is required on the workpiece with carbon dioxide MAGS spot welding, and a backing block is not required. Mismatch of the sheets is permissible with a maximum gap equivalent to half the sheet thickness, the extra metal being provided by the electrode wire. Up to 30 spot welds per minute may be made, which compares reasonably well with the 100 spots per minute from resistance welding techniques.

The deep penetration characteristics of carbon dioxide MAGS welding enable spot welding of widely differing metal thickness to be performed successfully, together with multisheet thickness.

Tungsten arc gas-shielded (TAGS) welding

This process (see Figure 87) was first used industrially in 1942 for joining magnesium alloys. It is more usually known as tungsten inert gas (TIG) welding and its application has extended to aluminium and its alloys and a wide range of ferrous alloys such as stainless steel, carbon steel and alloy steel to be welded without the use of a flux.

Principle of operation

The heat for welding is produced by an electric arc between the non-consumable tungsten electrode and the part to be welded. The heated weld zone and the molten metal and the tungsten electrode are shielded from the atmosphere by a blanket of inert gas (argon or helium)

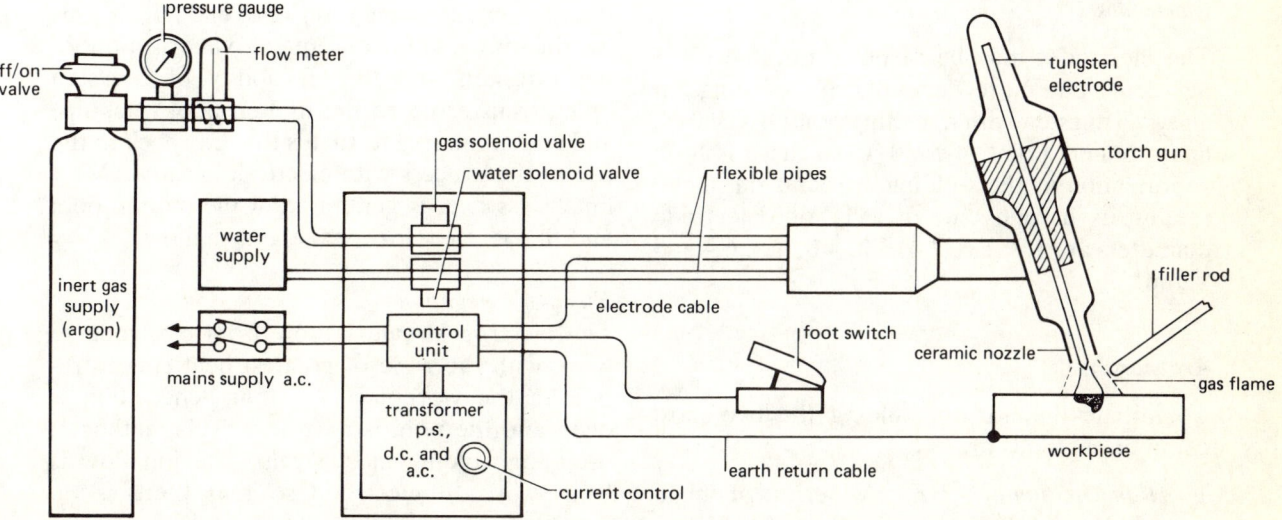

Figure 87 *Tungsten arc gas-shielded (TAGS) welding (tungsten inert gas or TIG)*

fed through the electrode holder. A weld is made by applying the arc so that the abutting workpieces and filler metal are melted and joined together as the weld solidifies; at the same time the inert gas shield excludes the atmosphere and prevents contamination of electrode and molten metal.

Gas shield

Argon in its commercial purity state (99.996 per cent) is used for the metals named but for titanium extreme purity is required. Argon with 5 per cent hydrogen gives increased welding speed and/or penetration in the welding of stainless steel and nickel alloys; nitrogen can be used for copper welding on deoxidized coppers only. Helium may be used for aluminium and its alloys and copper but it is more expensive than argon. Moreover, owing to its lower density, a greater volume is required than with argon to ensure adequate shielding and small variations in arc length cause greater changes in weld conditions so that manual welding with helium is not as easy as with argon. The mechanized d.c. welding of aluminium with helium gives deep penetration and high speeds.

Electrodes

The electrodes may be of pure tungsten (99.5 per cent) but more generally are of tungsten alloys. Tungsten with a melting point of 3380 °C has a boiling point of 5950 °C so there is little vaporization in the welding arc, and the metal retains its hardness when red hot. Electrode diameters are 1.2, 1.6, 2.4, 3.2, 4.0, 4.8, 6.4 and 8 mm.

Arc starting

To initiate the arc for welding the two most common methods are:

1 *High frequency (HF)* A series of high-voltage high-frequency sparks are superimposed on the main welding current so that, at the press of a switch, they pass from the tungsten to the work and so 'ionize' the air gap (break down the resistance) and allow the welding current to create an arc. This avoids touching the plate with the tungsten and avoids contamination. The HF may be continuous for a.c., and for d.c. is used only when the arc has been extinguished.
2 *Surge injection* This is another method of arc starting which also uses a high-voltage high-frequency spark.

Electrode polarity

Electrode positive
The electrode stream is from work to electrode, while the heavier and slower positive ions travel from electrode to workpiece. If the work is of aluminium or magnesium alloys there is always a thin layer of refractory oxide present over the surface, which in other processes has to be dispersed by means of a corrosive flux to ensure weldability. The positive ions in the plasma bombard this oxide and together with the electron emission from the plate break up and disperse the oxide film.

The electrons streaming to the tungsten electrode generate great heat; hence the electrode diameter must be relatively large, and a large blob is formed on the end. It is this overheating, with consequent vaporization of the tungsten and the possibility of tungsten being transferred to the molten pool (pick-up) and contaminating it, that is the drawback to the use of the process with electrode positive. Very much less heat is generated at the molten pool and this is, therefore, wide and shallow.

Electrode negative
The electron stream is now from electrode to work with the zone of greatest heat concentration in the workpiece, so that penetration is deep and the pool is narrower. This method is used for welding most steels. The ion flow is from work to electrode so that there is no dispersal of oxide film, and this polarity cannot be used for welding the light alloys. For a given

diameter, the electrode, when negative, will carry from four to eight times the current as when it is positive and twice as much as when a.c. is used.

Current

With alternating current, the advantage of positive electrode can be gained without the current limitations that are encountered with a positive electrode in d.c. welding. This advantage is offset to some extent by the need to provide means for maintenance of a stable arc. The use of a d.c. current, with electrode positive, overheats the electrode tip and tungsten inclusions can consequently be trapped in the weld. As a compromise alternating current is usually used for aluminium welding with a superimposed high-voltage surge or spark to assist reignition of the arc at each polarity reversal.

Characteristics

A clean weld of excellent metallurgical quality can be produced with precise control of penetration and weld shape in all positions, even in thin materials. As shielding is good, reactive metals such as aluminium and magnesium can be joined. Typical welding currents are low (less than 200 A) and, therefore, weld completion is slow.

Type of operation   manual mode.
Heat source   electric arc, usually electrode negative for steel, but electrode positive must be used for aluminium alloys
Shielding   inert gas – argon or helium.
Current range   10–300 A.

Metallic arc welding

This welding process is useful for welding the heavier gauge plates used in vehicle bodywork and also for the type of metal plate processes in which the metal ranges in thickness from 3.175 mm (⅛ in) upwards to 76.20 mm (3 in). In this process, sometimes referred to as 'stick' welding, the weld is made by an arc struck between an electrode and the work. The high-energy arc melts both the work and the electrode. The deposition of the weld metal is by the operator who is responsible for the correct fusion of the joint.

Basically a metal arc welding plant (Figure 88) consists of the following items:

1 The power source, which may be alternating current (a.c.) or direct current (d.c.).
2 The welding lead cable and electrode holder.
3 The welding return cable (*not* earth lead) and clamp.
4 The welding earth.

Two types of welding current are used: a.c., which changes from negative to positive at the frequency of the supply; and d.c., which flows in one direction only.

The a.c. 'mains' supply is not suitable for welding because the voltage is too high and the

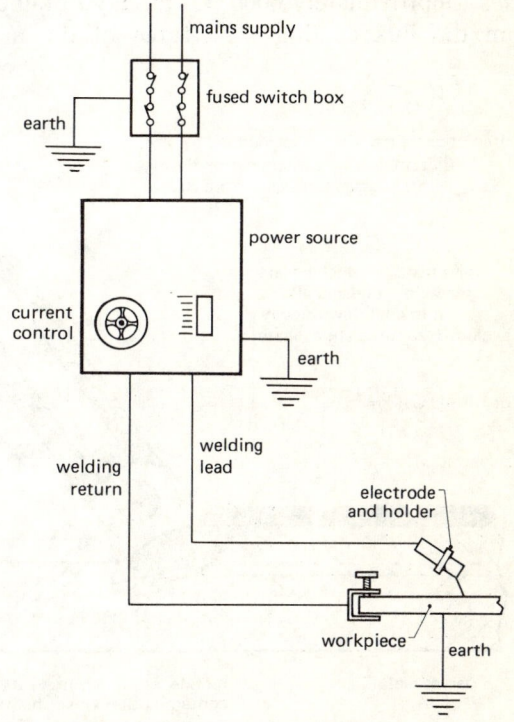

Figure 88   *Manual metal arc welding*

current too low. A transformer is used to change this to suitable values for welding, i.e. low voltage and a high amperage, but still alternating, current.

The electric arc and functions of the electrode coating

When the welding circuit is 'closed', by striking the electrode on to the work and withdrawing it slightly, an arc is formed (Figure 89). This contact is to enable a flow of current in the form of electrons to take place after the initially high voltage has overcome the resistance to current flow (sometimes called ionization of the arc gap).

The arc causes the parent metal to melt or fuse. The metal core of the electrode conveys the electrical energy to the arc and is melted along with the flux coating to form molten droplets of metal and flux. The arc is now composed of regions of very high temperature gases (approximately 6000 °C) mainly obtained from the flux coating. The force of the arc,

Table 3 Current selection

Electrode diameter (mm)	Length (mm)	Current (A)
2	300	50–80
2.5	350	60–95
3.25	350	80–130
3.25	450	80–125
4	350	120–190
4	450	120–180
5	450	150–250
6	450	190–310
6.3	450	200–370
8	450	300–500

helped by gravity and surface tension, project the molten droplets into the weld pool where they solidify under the protective covering of the solidified flux, now called slag. The flux also provides a shield of gas which protects the molten metal at both the electrode tip and the molten weld pool. In addition the flux supplies salts which provide ionized particles to assist

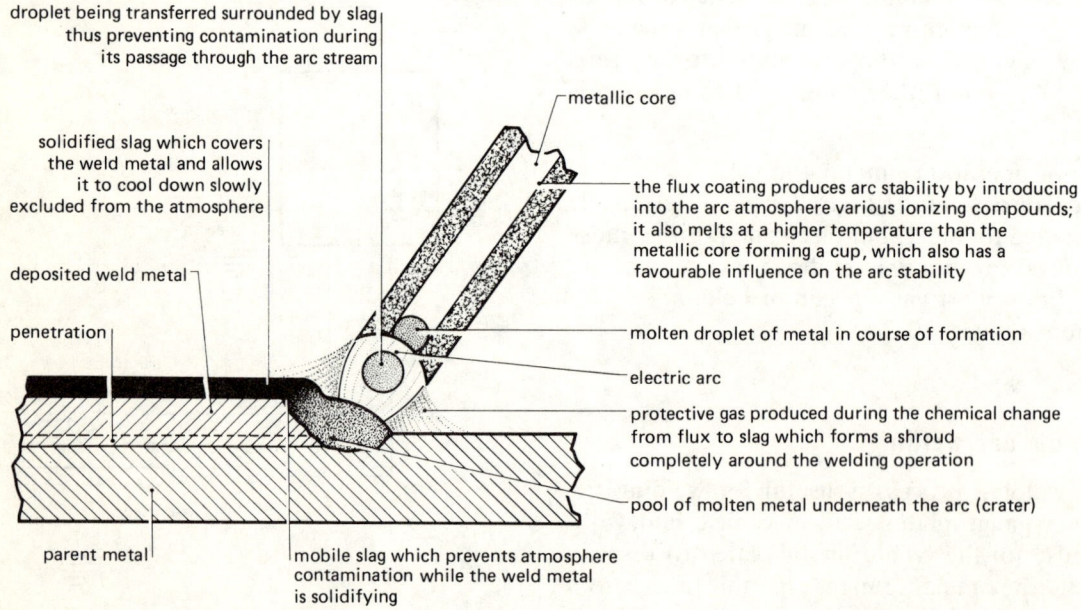

Figure 89 *The metallic arc: functions of the coating*

reignition of the a.c. arc. Sizes of stick electrodes are specified by the nominal diameter of the electrode core wire; the range available is shown in Table 3.

Variation in current values

Current too high
If the welding current is too high, a flat wide bead with coarse ripples results from the increased arc force. The arc force also produces a deep penetration pattern accompanied by an excessive amount of spatter. Though striking the arc is made easier, the crater is deep with blowholes and sometimes has cracks in the centre. Porosity (gas entrapped in the weld) is sometimes the result of too high a current. Overheating, particularly of thin-section material, and also of the electrode, results in unsatisfactory welded joints.

Current too low
If the welding current is too low, the arc is difficult to control and often the electrode end fuses to the plate. This causes a short circuit and the electrode becomes red hot due to resistance heating unless it is broken off. The weld bead tends to be high and globular and of irregular width, with slag trapped in crevices and difficult to remove. The penetration is shallow in the centre of the bead whilst the toes of the weld are often just adhered to the plate.

5 Collision damage repair equipment

Development

The use of hand tools alone cannot always correct some types of bodywork damage so other methods have to be used. In collision repair work, hydraulic body jacks along with hand tools and sometimes heat have to be used to realign and finish the damaged sections to their original designed shape. In collision damage the body outer panel is distorted and this more than likely involves some damage to the inner panels or structure as well. When correcting the damage, both inner construction and the outer panel must be straightened and aligned at the same time unless of course the outer panel has to be removed for replacement.

The use of hydraulic body jacking equipment has been developed to its present sophisticated state over many years. In the early days the ratchet or screw-type body jack was the only type available, but it had many disadvantages. The power that could be developed was small and would certainly be inadequate to cope with correcting crash damage on present-day cars.

Mechanical equipment was limited in scope as well as in power and much of the effort put in by the operator was lost in the friction of the screw or ratchet. This in turn reduced the 'feel' of the operator and tended to make accurate control of the job more difficult. The ratchet-type jack worked in steps instead of giving the smooth, controlled advance of the hydraulic body jack and this also made control more difficult. The screw-type jack tended to apply a torque or twisting motion to the pusher head instead of providing a straight push. All mechanical equipment had to be operated from within the car, whereas the hydraulic body jack is remotely controlled from outside the vehicle, allowing the operator to be in the best possible place to see the job as it progresses.

The use of hydraulic body jacking equipment grew from the use of an ordinary hydraulic car jack for this purpose. The hydraulic hand jack had all the advantages of providing tonnes of closely controlled torque-free power for the minimum of operator effort, but was limited to straightforward pushing in a vertical or horizontal plane. It soon became apparent that what was required was a hydraulic jack that could be operated in any plane and controlled from outside the car. This was achieved by separating the cylinder ram part of the jack from the pumping unit and then connecting them with a flexible hose.

Porto-Power

The resulting Porto-Power hydraulic unit has remained basically unchanged for many years. The pump unit comprises a reservoir, pump, handle and hose, and is controlled by a simple 'open and close' release valve (Figure 90). The handle can be screwed into the pump in two different positions for ease of operation. Alternatively, for light work, the unit can be controlled more finely by dispensing with the handle and pumping by means of the beam only.

The hose is connected to the ram by a simple quick-release coupler which needs to be done up only finger tight. Like this, it can still swivel to allow the equipment to be got into position and to get twists out of the hose, but when pressure is applied it tightens up completely to prevent

Collision damage repair equipment 79

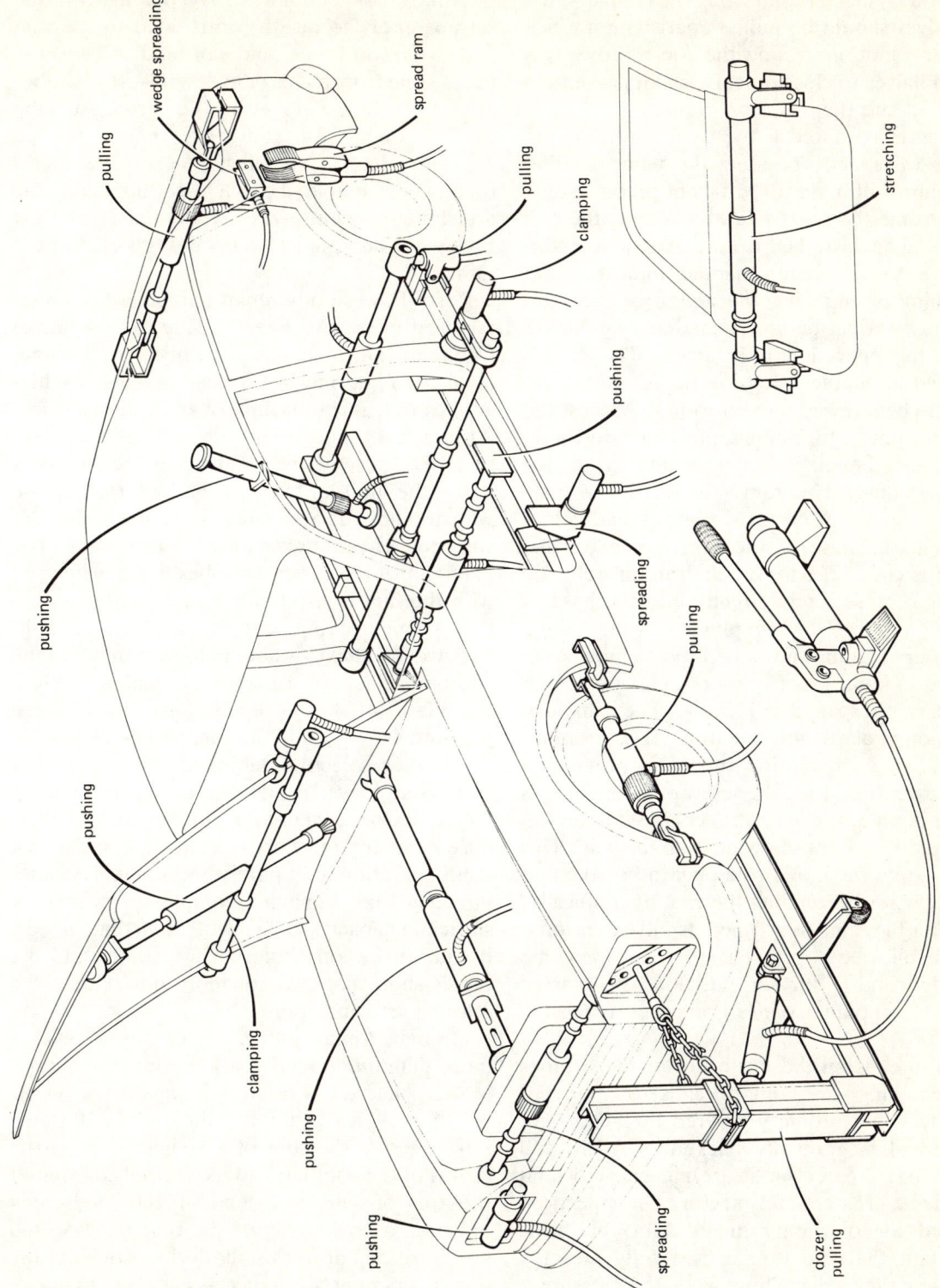

Figure 90 *Hydraulic jacking equipment*

loss of oil. When disconnecting the couplers, the two halves should be pulled apart with a quick 'snapping' motion to allow the non-return valves in both halves to close quickly, again preventing loss of oil and the inclusion of air.

Pressure is applied by closing the pump release valve and operating the pump handle. The pump will build up sufficient pressure only to overcome the resistance of whatever the ram is pushing against. The need to apply excessive pressure to the pump handle indicates that something is impeding the action of the ram. The cause should be ascertained and dealt with before the operation is resumed. The ram is threaded to enable extension tubes and attachments to be screwed into position to harness the hydraulic power for pulling, pushing, stretching, tensioning, bending, straightening, clamping and spreading. The ram plunger has a male tapered thread. A corresponding thread is in a recess in the base of the ram and a parallel thread is cut at the top of the ram barrel. The extension tubes screw directly into the base of the ram, or on to the ram plunger by means of a tube coupling, and the same tube couplings can be used to join tubes together to give the length necessary to span a car body. The slip lock extension is a valuable addition. It screws into the end of an extension tube and provides a screw extension. Like the screw of an engineer's vice, it has a quick release to enable it to be run out rapidly without screwing it all the way. This screw extension allows free play to be taken up quickly before extending the ram hydraulically, and after the ram has reached its full extension it can be allowed to retract, the screw can be extended and a further push can be started without stopping to insert another extension tube.

The threads on the Porto-Power rams, tubes and coupling are full threads on tapered diameters, not ordinary tapered threads where the thread is gradually backed off. This full thread has greater load-carrying capacity and provides a greater safety factor. Units need be screwed together only finger tight, but it is important that the threads are fully engaged otherwise they will collapse when the pressure is put on. It must be remembered that many of the set-ups apply to an off-centre load to the ram and extension tubes, and can tend to bend the tubes unless work-hardened solid drawn tubes are used; however, even the strongest tube made cannot withstand more than a limited off-centre load. Care must be exercised therefore when employing such set-ups and the attachments themselves are designed for light bodywork only, not to carry the full load of the ram.

Porto-Power body repair equipment is manufactured in various sizes ranging from 4 tonne, for light vehicle bodywork only, to 20 tonne, ideal for repairing heavy road-making machinery, trucks, farm machinery and railway stock. Although 4 tonne would probably be sufficient to move most of the reinforcing sections of a body, the limiting factor is the load that can be withstood by the attachments of a size consistent with the physical size of a 4 tonne ram. The diameter of the extension tubes that can be used with the ram is also limited, and for this reason the 4 tonne equipment is suitable for light bodywork only. This does not mean that 4 tonne equipment is not capable of dealing with a multitude of body repair jobs on its own account. However, limitations should be kept in mind when selecting equipment. Being smaller and lighter in weight, it is less expensive than the larger 10 tonne standard equipment, but it is false economy to select 4 tonne equipment as a main operational set purely because it is cheaper initially. Like buying a compressor with only sufficient capacity to deal with immediate needs, it will not be long before a job comes into the repair shop that calls for more power than the 4 tonne set can supply.

On many repair jobs it is necessary to push in more than one direction at the same time and it is here that the 4 tonne equipment proves its worth. A typical example is the use of a 10 tonne set-up across the roof of a vehicle to raise the crown of the roof back to its original contour. If the area of damage extends to the windscreen opening, a tension across the roof panel would not be sufficient to raise both the panel and the screen opening. It is necessary then to use a

4 tonne set-up in conjunction with the standard equipment. With the 10 tonne set-up applying the tension to the roof panel, the 4 tonne equipment is used as an auxiliary set pushing directly upwards under the damaged section of the screen opening. The 4 tonne set is carrying the lighter of the two loads and the two together achieve the desired result.

The 10 tonne standard equipment is the mainstay of the body repair trade and the range of attachments and special-purpose rams covers all aspects of body repair work. The set contains the basic essentials for body repairs and from it can be built up gradually a more comprehensive set until it is possible to deal with the more complicated repairs.

The limitations of Porto-Power equipment can be summed up under three headings:

1. Difficulty in finding somewhere strong enough to push from. Basically the ordinary Porto-Power pushes from one point only, and although it is possible to spread the load to a limited extent by using a base plate on the ram or extension tube, or a baulk of timber, it is still not possible to get away from the limitation of single-point pushing.
2. Restricted direction of thrust. Assuming that it has been possible to find a single point strong enough to push from, the chances are remote that this point will be exactly in line with the direction in which it is desired to push. This makes necessary the use of complicated chain anchorage set-ups to restrict the direction of thrust to the required line.
3. Intervening obstructions. Frequently, the line between the point from which the equipment is pushing and the point at which the load is being applied is obstructed by a major mechanical component like the engine which has to be removed just to give access for the equipment. At best a considerable amount of interior trim has to be removed, all of which adds to the cost of the job without contributing directly to the repair.

These limitations influenced the introduction of the now firmly established Dozer technique.

The Dozer technique

The method of rectifying damage by pulling from the outside is not new; in many body shops a steel ring had been fixed to the floor or to an upright beam set in the concrete floor and this had been used as an anchorage for pulling from. This was a hit and miss method at best, and in some cases it has been known for a damaged car to be shackled to such an anchorage and either rocked, towed or even driven backwards to achieve an outwards pull. There was little accuracy or control with such a haphazard operation.

Alternatively, a heavy steel 'bedstead' is used and it is necessary to employ lifting tackle to lift the vehicle bodily to mount it on the rig. The installation of such a rig and the reinforcing of the shop superstructure to support the lifting tackle is very costly, and even when completed, only one work bay is available for this type of major repair. Once a vehicle has been put in position it cannot be removed until all the bodywork has been completed and in addition until the mechanical work to the suspension, etc. has been completed. Thus other major body repairs may be held up whilst the previous car occupies the rig until it can be rebuilt sufficiently to stand on its own wheels again. A portable pulling frame was developed that could be taken to the car in any part of the body shop and could be moved onto the next job once straightening had been completed. This frame was called a 'Dozer' and several different types were designed to cater for the various classes of repair work.

Basically the Dozer (Figure 91) consists of a base beam, mounted on casters for ease of movement, with an upright beam pivoted to it at one end. Across the angle between the two beams a hydraulic push ram is mounted so that when the ram is extended the upright beam moves in an arc backwards. The free end of the base beam is anchored to more than one strong point on the car underframe by special clamps and chains and the damaged area is shackled to the upright beam. Then when the pump is operated the ram moves the upright beam

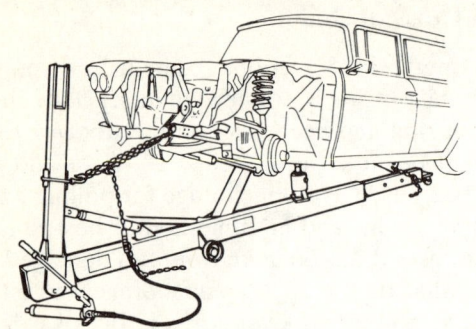

Figure 91 *Portable dozer*

backwards with controlled pressure pulling out the damaged section under complete and precise control. In this way the three main limitations of conventional Porto-Power have been overcome:

1. By taking the reaction to more than one of the strongest points on the car underframe, the problem of a single point strong enough to push from is overcome.
2. As the Dozer can be lined up to pull in any direction, the problem of direction of pull does not arise.
3. As all connections are made externally there is no necessity to strip interior trim or remove assemblies to provide access.

The first Dozer designed was the Pull-Dozer, a small frame about 2 m long and 1.2 m high, powered by a 4 tonne hydraulic unit and intended for light panel and sheet metal work only. Clamps and chains are provided for efficient anchorage, but these are not engineered for work on chassis or reinforced sections. It is possible to substitute the 10 tonne unit for the 4 tonne unit in the Pull-Dozer frame but it should be remembered that neither the frame nor the attachments are designed to take the full ram load under these circumstances. This facility is offered simply so that body shops already using standard Porto-Power can utilize their existing pump and ram in the Pull-Dozer frame. However, as the Pull-Dozer is not suitable for correcting frame or chassis damage, it is often necessary to use Standard Porto-Power in conjunction with the Pull-Dozer to correct frame and panel damage in one operation; hence it is preferable to have a separate hydraulic unit for the Pull-Dozer and not to try to make one unit do for two.

A typical example of the combined use of Dozer and Porto-Power is the repair of a rear-end collision where the rear quarter panel has concertinad and the reinforcing member over the wheel arch has been forced upwards. The Pull-Dozer is positioned at the rear to pull out the crumpled panel and at the same time the standard 10 tonne ram with extension tubes and pusher heads is placed across the spring hangers to return the wheel arch member to its proper position.

For complete body and frame straightening in one operation with one piece of equipment, the Unit-Dozer or Damage-Dozer was developed. The former is basically the same as the Pull-Dozer except that it is larger and powered by a 10 tonne unit. The chains and attachments are correspondingly heavier and the method of anchorage to the car is modified to take the greater load imposed. The clamps are of a different design to those supplied with the Pull-Dozer and longitudinal adjustment is provided by means of an anchor post which slides backwards and forwards on the base beam. As with the Pull-Dozer, anchorage on the upright beam is provided by a sliding chain hook. The higher it is raised on the upright member, the smaller the force applied, but the greater the length of pull.

Conversely with a lower anchorage a shorter stronger pull is obtained. If the beam anchorage is lower than the point of pull on the car body, then a downward pull is obtained and vice versa. In the first case some support has to be inserted between the Dozer base beam and the underside of the car to prevent the Dozer from listing. In the second case a chain anchorage is necessary to shackle the car down to the Dozer beam.

The type of body clamp used to anchor the rear end of the base beam to the car body is known as the unitized body clamp. The pair of clamps bite into the pinch weld running the length of the body on each side of the car from front to rear wheel arches. A support tube passes through the bosses in the body clamps, and a reinforcing tube slides over the top of the support tube between the body clamps. The ends of the support protrude beyond the width of the car body and are located in saddles of a pair of safety stands similar in operation to the ratchet-type axle stands. This set-up raises one end of the car clear of the garage floor and the other end should be supported by ordinary axle stands.

The raising of the car clear of the garage floor is important to provide access for the Dozer, the rear anchor post of which butts up against the support tube running the width of the body. Also, the car is now raised clear of any damaged suspension members which may have given a false impression of frame distortion if the car had been checked for alignment whilst standing on its own wheels. A firm grip on the frame or panel member to be pulled out is obtained by a clamp set which has a pin riding in a taper so that the grip of the jaws becomes tighter as the load applied becomes greater.

The Damage-Dozer is similar to the Unit-Dozer except that longitudinal adjustment is by means of a telescopic base beam instead of a sliding rear anchor post. The base beam is supplemented by a cross-beam which can be anchored in position to provide a cruciform giving four points of anchorage for removing twist from a body. It is secured by a large U bolt and clamp plate and can readily be removed or replaced as required.

Operation of conventional Porto-Power

All collision damage repair is carried out by using one or other of a number of simple set-ups, or in the case of more complicated repairs a combination of set-ups. The first important step therefore is to understand the set-ups that can be built up, the attachments required and their application. After this it is a question of breaking down a job into its basic set-ups and applying the corrective force in the correct sequence.

Pushing

Pushing (Figure 92) is the simplest operation of all and is achieved by inserting the ram between two points and operating the pump. The plunger extends until it touches the point at which the load is to be applied and as pumping is continued pressure is built up to overcome the resistance of the metal at the point of application.

Movement of the damaged area will continue as long as pumping is carried out. It must be remembered that a greater force is required to straighten a damaged section than was needed to distort it. Therefore, if a ram was placed, say, between two underbody members and it was intended to push from the undamaged member to straighten the damaged one, the force applied would not rectify the damage but would distort the undamaged member. From this it will be seen that the first essential is to ensure that the point you are pushing from is stronger than the

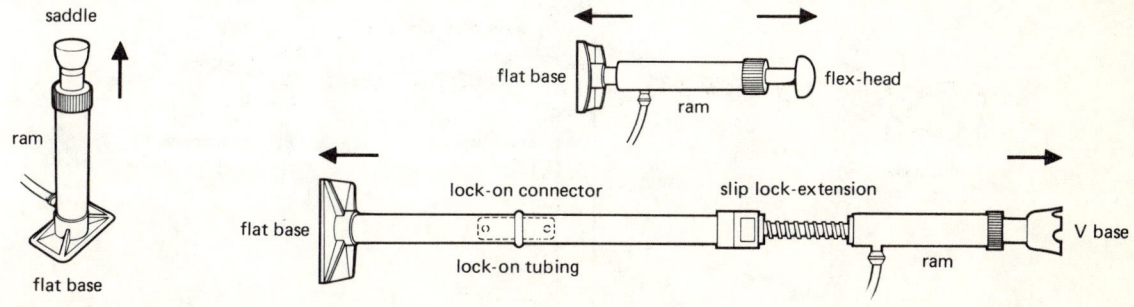

Figure 92 *Pushing set-ups*

point at which you are applying corrective force. This can be done by attaching a base plate to the bottom end of the ram to spread the load and the load can be spread over an even larger area by interposing a piece of timber between the base plate and the pushing point. It is seldom that a pushing application can be achieved using the ram only, and various combinations of extension tubes couplings, pusher heads, etc. are required to cater for any job requiring a straightforward push. Examples include the base plate, saddle, V-shaped saddle and flex heads. Flex heads are made of hard rubber with threaded metal inserts for attaching them to the ram or extension tubes. They conform to the contour of the area against which they are pushing and obviate the risk of damage to the panel that might occur with a metal head. Being rubber they will not slip easily if it is necessary to push at an angle but care should be taken to see that the flex head is not pushing against the edge of a pinch weld or similar sharp section as this would cut into the rubber. Sometimes it may be necessary to accept minor damage of this nature to a component like a flex head for the sake of getting a job done, in which case the item can be regarded as semi-expendable and can be replaced relatively cheaply when its useful life has been exceeded.

Pulling

Pulling (Figure 93) is also a simple operation but uses a slightly different type of ram. With the standard Porto-Power a pull converter set is required to enable the one push ram to fulfil both functions, but it gives an off-centre pull and is less convenient to use than the separate pull ram. With direct pulling set-ups using the pull rams, the tubes are under tension and there is no risk of bending. The thread sections of tubes and couplings are now under tension and it is the threads themselves (not the taper, as with pushing) that carry the load, so again it is important to ensure that threads are kept free from dirt and damage and are screwed up properly.

The pulling combinations make use of the pull rams and direct pulling attachments, although it is possible to use chain plates and chains for obtaining a pull with a push ram. The latter method is quite convenient if, for example, you wish to pull across the width of a body from an undamaged door pillar to a door pillar which has been bowed outwards. It would be necessary to reinforce the undamaged pillar with timber to prevent its distortion and to protect both pillars from scuffing by chains with suitable packing material.

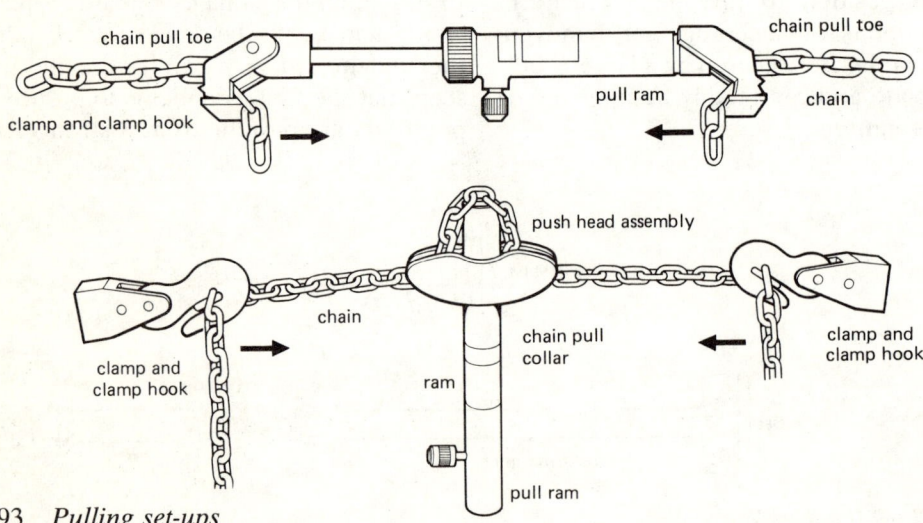

Figure 93 *Pulling set-ups*

Collision damage repair equipment

Spreading

Spreading (Figure 94) is a similar application to pushing except that in the latter case there is sufficient room for access for the ram and extension between the two members to be moved apart to permit a direct push. When this is not possible, a means must be found of inserting jaws or attachments which are capable of applying an indirect thrust. The most obvious means of spreading is provided by the 'wedgies' and spread rams. However, because of their mechanical linkage, which is necessary to transform a small hydraulic movement into a much greater movement of the jaws, the amount of tonnage that they can apply is strictly limited. However, by using a toe screwed on to the ram barrel and another toe or base plate on the plunger, a spreading combination is achieved with the ordinary push ram. This is an off-centre load once more, and even under the most favourable conditions it is not possible to apply more than about 7 tonne with the 10 tonne ram or 2.5 tonne with the 4 tonne ram, but this is considerably in excess of the tonnage obtainable with the spread rams.

Stretching or tensioning

The technique of stretching or tensioning (Figure 95) is another means of obtaining a pull using a push ram. An external pull is applied to pull apart or draw outwards areas that have been pushed or drawn in towards each other. The combinations used to obtain this external pull employ toes, links and clamps on the end of the push ram and extension tubes. Two types of clamps are the pull clamps for attaching to flat edges and the fender clamps which have deep throats for lipped edges. Both clamps have an alligator type of action and are first tightened down on the centre bolt until the jaws are parallel and in contact with the surface of the panel to be gripped. Pressure is then applied by tightening the rear bolt which cants the jaws forward and causes them to bite into the surface of the metal. It is essential that all paint must first be removed from the surface of the panel where the jaws are to bite if a satisfactory grip is to be obtained. Overtightening of the rear screw will break the clamp jaw across the line of the centre screw.

On flat panels where it is not possible to get at

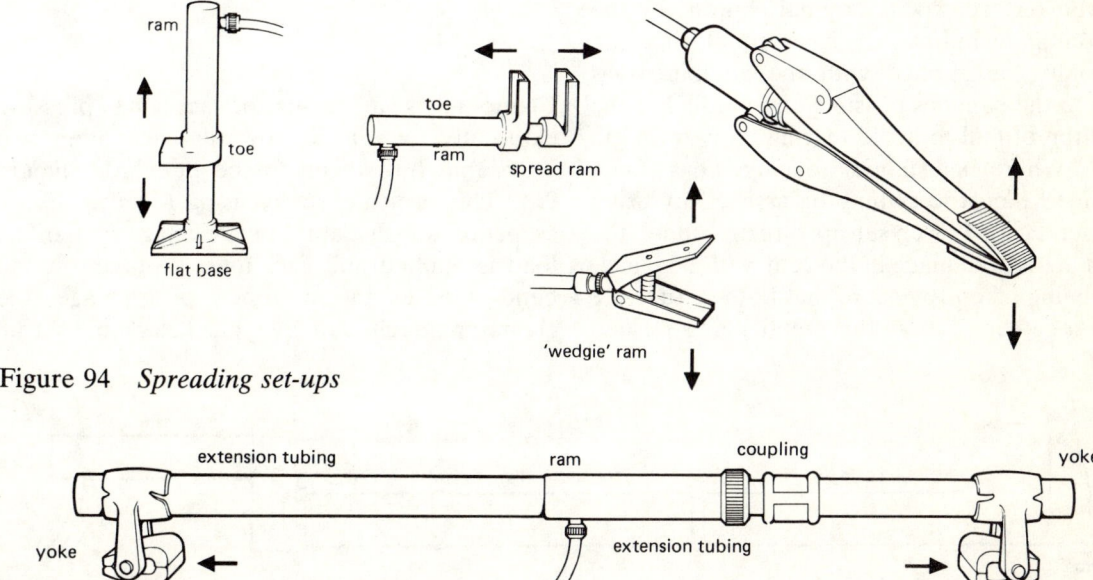

Figure 94 *Spreading set-ups*

Figure 95 *Stretching or tensioning set-up*

the edge or where there is no edge to locate a clamp, the same result can be achieved by locating the toes in the bosses of a pair of solder plates which have been sweated on. A roof panel or a car door are typical examples of this. Usually to get at both edges of a car door it is necessary to remove the door, but this non-productive time can be eliminated by using a clamp at the free edge of the door and a solder plate at the hinged edge. The plate should be sweated on with a 3 mm (⅛ in) thick layer of body solder using the minimum of heat necessary to melt the solder and quenching with a wet rag to prevent the spread of heat through the panel. In this way there will be no risk of further distorting the panel through the application of too much heat.

If a roof panel, for example, has been dented in, the edges of the panel will have been drawn into the centre. If pressure was applied directly under the damaged area to push it back up further damage would occur. It is necessary to apply a pull across the panel drawing the edges outwards to their original position and raising the crown of the roof at the same time. This is done in practice by sweating solder plates or metal strips on to the roof or other panel which is to be restored to its original contour by the tensioning technique. It is essential that the tensioning be applied with the equipment as close to the panel as possible and parallel to the direction of pull to avoid tearing away from the roof. Where substantial damage has been sustained the process may be assisted by using another Porto-Power set-up directly under the worst areas of damage in the centre of the panel, care being taken to ensure that both set-ups are used together so that the metal being pushed directly upwards can move outwards at the same time.

Where the damage to the panel extends into low-crown areas that have been made rigid by forming during manufacture, or into door, window or windscreen openings, the pressure on the crown of the roof resulting from these areas of secondary damage will restrict the movement of the panel unless they are dealt with at the same time. In such cases 4 tonne equipment is most useful for working on these secondary areas whilst 10 tonne equipment is employed on tensioning the panel itself.

Finally the panel beater will use the hammer and dolly to iron out creases on the panel before releasing pressure. The whole process is one of gradually working from the outside in towards the centre, applying the tension, relieving restricting secondary damage, plannishing to remove the creases, then repeating the whole operation a stage further, until the final result is achieved.

The same technique applied to door panels, boot and bonnet lids, front wings and rear quarter panels, can be the means of repairing a panel *in situ*, whereas direct pressure from inside would not have achieved a satisfactory result.

Clamping

Clamping is a means of applying pressure externally to bring in towards each other two areas that have been forced outwards (Figure 96). This is achieved by using a set-up like a carpenter's sash clamp, but once again an offset load is applied and care must be taken not to bend the extension tubes or damage the clamping attachment. A typical clamping set-up

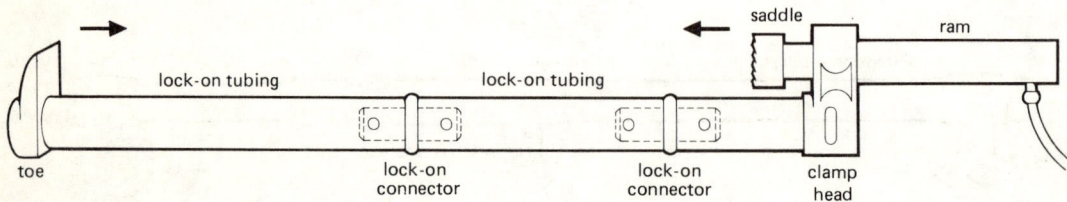

Figure 96 *Clamping set-up*

is the restoring of a bowed-out door pillar by clamping across the width of the body from the undamaged door pillar reinforced with a piece of timber.

Pressing

Pressing (as opposed to pushing) does not enter very much into the field of collision damage repairs, but mention is made of the portable press for the removal of king pins etc., as the stripping out of suspensions for rebuilding or even to gain access for repairing body damage is sometimes a feature of repair work. Similarly the straightening of components such as bumpers, which would normally have to be removed from the car in any case, can be carried out in a small bench press.

Operation of Dozers

The following paragraphs deal with Dozer set-ups and the results that can be achieved with them. Reference will be made to frame misalignment and it should be kept in mind that this refers not only to the chassis frame on the type of car with separate body and frame, but also to the reinforcing members that serve as a frame on cars of unit or monocoque construction.

A method of checking for frame misalignment can be carried out using a set of three frame gauges which can be mounted in various ways. They are mounted by means of adjustable hangers, one at the front, one at the rear and one half-way along the car. Each hanger must be located at a point symmetrically disposed about the centre line of the car (see Figure 97).

If the frame is in alignment all three gauges will be parallel and the pointers in line. If it is out of parallel this indicates that a section of the frame has been moved up or down by the impact, and if the pointers are out of line this shows movement to one side or the other. The gauges may be left in position as corrective work proceeds as an indication of the progress of the work. Overcorrection may be necessary to take care of natural spring in the body shell and this too may be judged by pulling into line, letting off pressure, observing how far the pins move back out of line, then applying the same amount of overcorrection. It is necessary also to use a diamond detector gauge to detect the condition when the frame has been pushed into a diamond or parallelogram shape (Figure 98). This is not a frequent condition but is one that would not be revealed by the frame gauges alone.

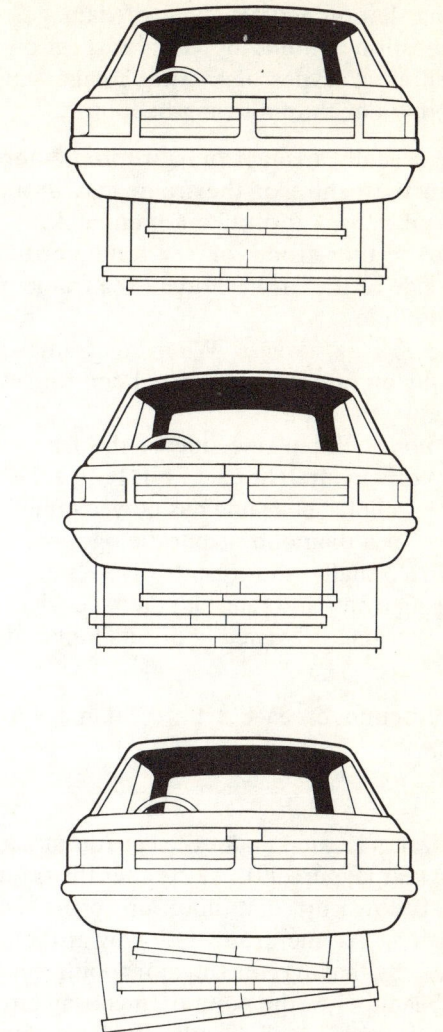

Figure 97 *Frame gauges showing distortion*

Having determined the type of damage, the actual location is found by visual inspection for damaged area. Types of frame damage can be classified under the following headings:

1. *Sag* Usually arising from a front-end impact when the centre of the frame has dropped and takes up a form like a hammock.
2. *Sway* Either front- or rear-end impact to one side of the car has forced the frame over to one side.
3. *Kick-up (or down)* When a frame leg, usually on one side only, has been forced up or down by impact.
4. *Diamond* When one side of the frame has been driven straight back, but both sides are still parallel. The frame has now assumed the shape of a diamond or parallelogram.
5. *Twist* Usually the result of a 'sandwich' causing front- and rear-end damage when the whole frame is twisted about its longitudinal axis.

The correction of each type of damage is as follows:

Sag
The Dozer is located under the car running from front to rear (Figure 99). At the rear the frame is anchored down by chain hook-up to the Dozer base beam. At the front a downward pull is obtained by having the anchor point on the swivel beam below the point of anchorage to the car front end. A hydraulic jack is placed between the Dozer base beam and the frame at the point of sag. By pulling down at the front and pushing up at the centre, the sag is eliminated.

Sway
When the frame has swayed to one side from the centre forward, the condition is corrected by positioning the Dozer across the vehicle and anchoring the centre of the frame by a double chain hook-up to the pivot beam (Figure 100). The rear anchor post of the Dozer is attached to either end of the frame and a centre pull is applied to eliminate the sway.

Figure 98 *Diamond shape frame*

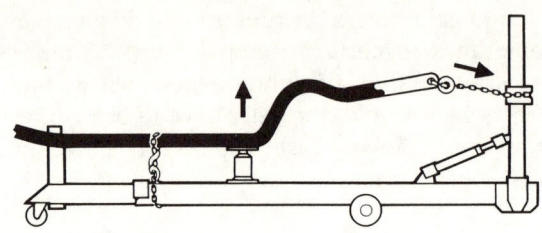

Figure 99 *Sag*

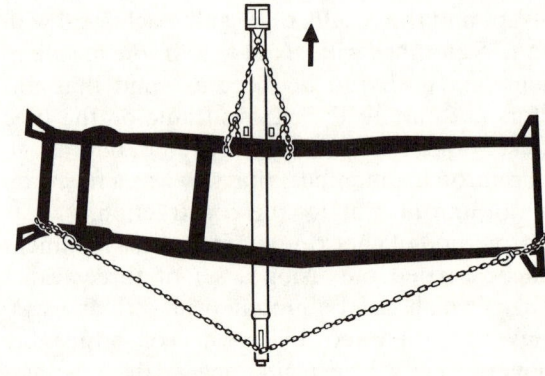

Figure 100 *Sway*

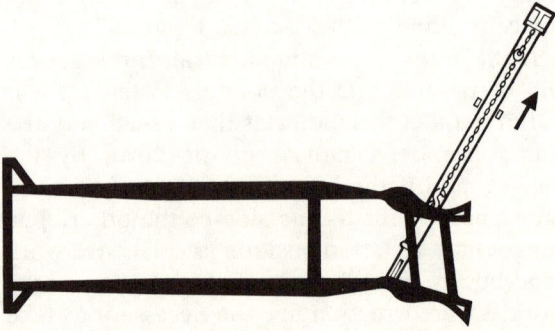

Figure 101 *Sway*

If however the sway has occurred at one end of the frame only, another type of hook-up is used (Figure 101). The Dozer then is set at an angle to provide an oblique pull with rear anchor post set at the point on the frame from which the sway has started. Thus when pressure is applied the frame forward of the anchor post is pulled across until it is realigned.

Kick-up

The hook-up is similar to the one used to correct sag, except that it is no longer necessary to push up at the frame centre with the hydraulic jack (Figure 102). The jack is still used to take up the space between frame and Dozer base beam, to prevent the Dozer from lifting when the pull is applied, but the only force used is a down pull at the front. If the frame has been kicked down, a similar set-up is used but the front anchorage on the pivot post is set above the point of attachment to the car, and the frame centre is held down by a chain around the Dozer base beam. When pressure is applied an upwards pull is obtained. Alternatively the pivot post anchorage is not used, and with the frame secured to the base beam at the centre point a direct upwards push is used with the jack or Porto-Power unit located between the Dozer base beam and the front frame member.

Diamond

A diamond condition is corrected by placing the Dozer diagonally across the frame on the shortest diagonal (Figure 103). The rear anchor post locates in the rear angle of the frame and the front corner is shackled to the pivot beam. A direct pull elongates the shorter diagonal until the frame is restored to correct alignment.

Twist

The removal of twist requires the use of the Damage-Dozer, whereas all the previous types of damage can be corrected using either the Unit- or the Damage-Dozer. The twist beam is put in position on the Dozer base beam and the Dozer is located under the frame with the two low corners fore and aft of the base beam

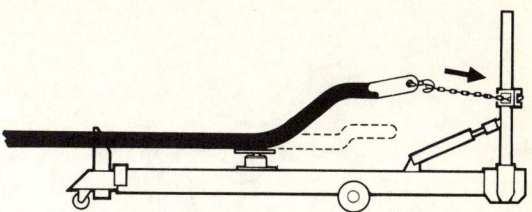
Figure 102 *Kick-up*

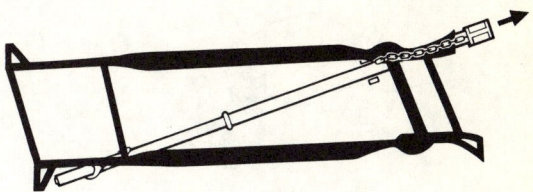
Figure 103 *Diamond correction*

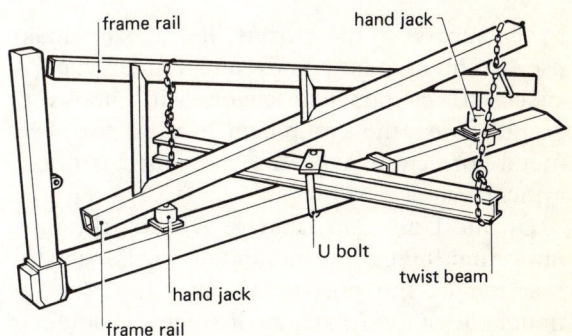
Figure 104 *Twist*

(Figure 104). The two high corners will now be located above the ends of the cross-beam to which they will be anchored by holding-down chains. The low corners are now raised with hydraulic jacks or Porto-Power units. If necessary, pull rams may be inserted in the holding-down chain linkages to obtain a downward pull in addition to the upthrust at the low corners.

Anchorages

The various methods of anchorage that have been contrived make an important contribution

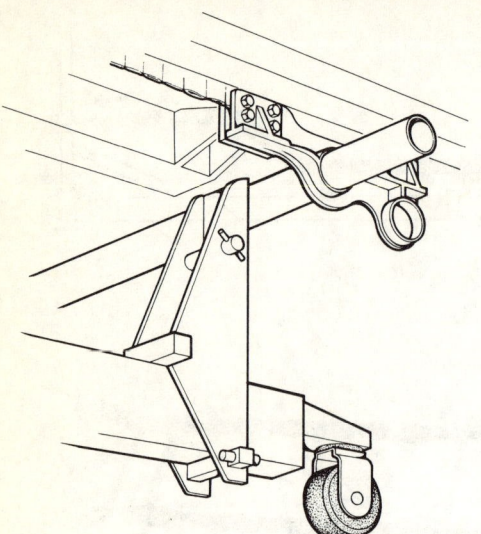

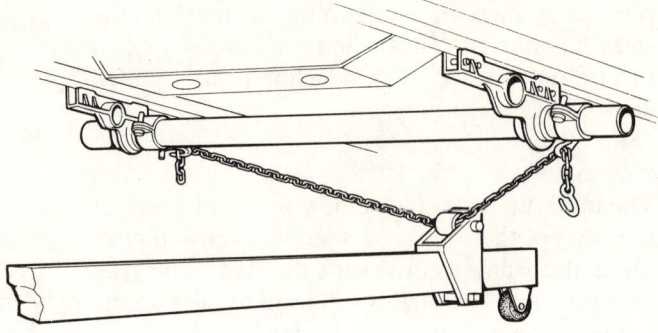

Figure 105 *Dozer anchorage*

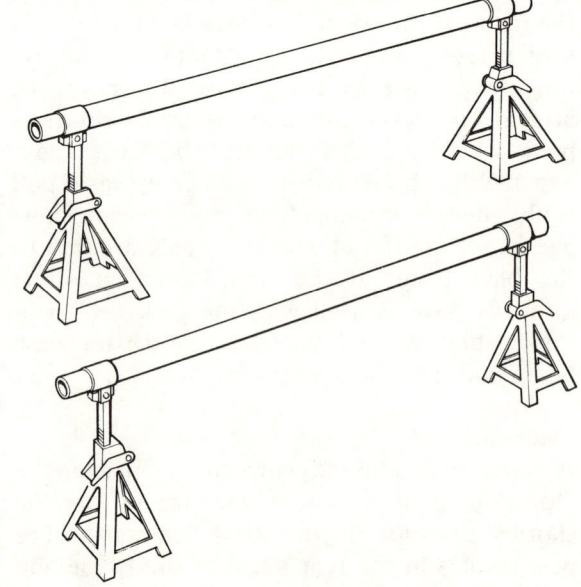

Figure 106 *Safety stands*

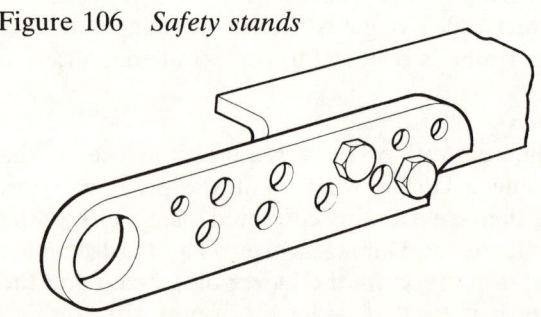

Figure 107 *Pull plate*

to the success of the Dozers. Pull-Dozers make use of the standard body and fender clamps, special C clamps and chains with hooks or grapples. As the equipment is used for sheet metal correction only, the anchorage is not so critical.

On the Unit- and Damage-Dozers, the most important thing is the method of transferring the reaction of the pull to strong points on the underside of the frame. A clear view is obtained of a body clamp in position on the pinch weld (Figure 105). With one such clamp in position on either side of the car, a support tube is passed through the bosses in the clamps with the addition of a reinforcing tube for greater strength. The rear anchor post of the Dozer butts against the reinforcing tube and the ends of the support tube are held by the adjustable safety stands (Figure 106).

Having obtained a firm anchorage for the Dozer itself, the remaining problem is to hook up the damaged part to the front pivot post to obtain the desired pull.

The first type of anchorage is provided by the pull plate (Figure 107) which has assorted holes for bolting on to such places as bumper mounting points and has provision for linking to the pivot post by chains.

The second type of anchorage is by means of a self-tightening clamp (Figure 108). The body of the clamp is a yoke which has a boss for passing the link chain through. The clamp jaw is tightened down parallel by a single screw, and is attached to the yoke by a pivot pin. This pin rides in the tapered rear portion of the jaws and the greater the pull exerted the more the front of the jaws closes giving a tighter grip.

Finally, chains form an important part of the anchorage system and with each Dozer special chains are supplied appropriate to the work for which the model is intended. The hooks on the chain ends are not conventional lifting hooks but are designed to hook back over the chain links to provide a firm hold. In addition to being used for a direct pull, the chains provide a holding-down link between Dozer and frame, and restrict the movement of the Dozer on oblique pulls.

The body bay system (Flexi-Force)

Many of the methods and much of the equipment in body repair are based on push rams inside the vehicle. However, such equipment creates many problems such as finding suitable locations from which to operate with the attendant risk of causing further damage. They also invariably involve removing seats and trims at considerable cost. Almost all the problems are overcome by using external pulling equipment and the body bay system comprises a number of steel sockets sunk into the floor of the crash repair bay which are the housings for steel posts (pull posts). These posts provide anchorage for the hydraulic pulling rams and by selecting the correct socket the post may be placed so that damage can be readily pulled out at the correct angle.

When the nature of the damage requires a diagonal pull (Figure 109), difficulties arise if conventional equipment is used, but with body bay pull posts sited both diagonally opposite and at the nearside rear (to prevent twisting) the

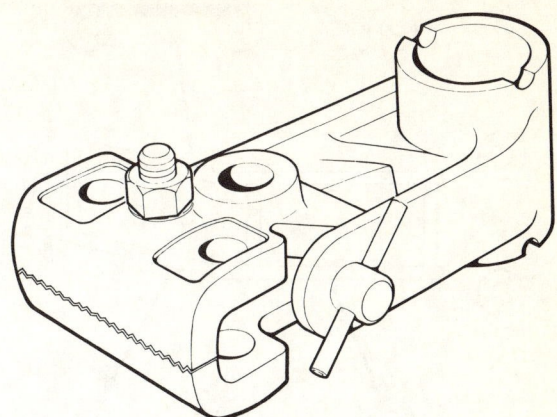

Figure 108 *Self-tightening clamp*

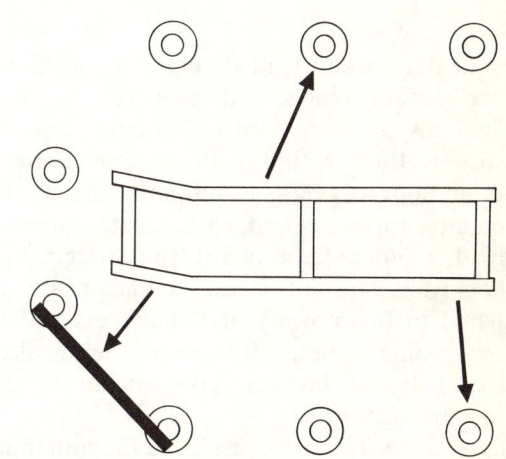

Figure 109 *Diagonal pull*

difficulties of the diagonal pull can be overcome. The power of the pull ram, which is attached to the pull posts, can then be applied and the damage easily dealt with.

Typical chassis damage (Figure 110) can be corrected by anchoring at two points to hold against the pull ram sill and floor damage can be corrected in a similar way.

A strong point often used for anchorage is the spring hanger bracket but this is not always necessary or available. By the use of a grip plate attached to the sill pinch weld, anchorage to

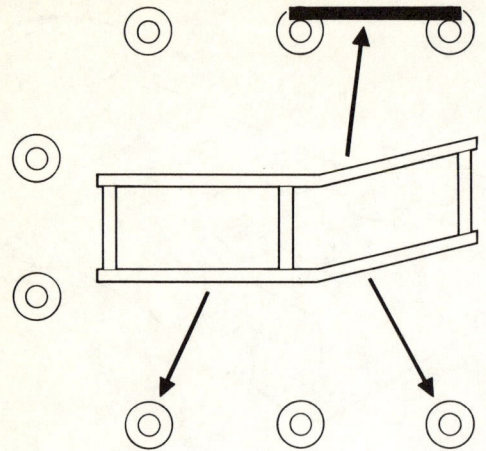

Figure 110   *Chassis damage*

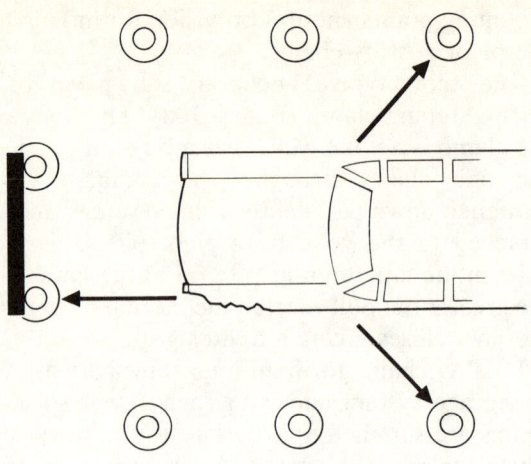

Figure 111   *Longitudinal pulling*

posts in side sockets is made possible and stress reduced on undamaged body parts (Figure 111).

The body bay is formed by installing a series of sockets flush with the floor, thus using a concrete floor as a basis to support either thrust or counter-thrust. Metal cover plates prevent body filler and underbody dirt from clogging the sockets which are not in use. Anchor posts are designed to fit into any of the sockets and to follow the line of pull. Prior to a pull being taken the car should be correctly anchored and pretensioned.

Six floor sockets are considered the minimum and it is recommended that they should be sited as in Figure 112(a). The side sockets allow floor and sill damage to be pulled, and while a major job occupies the central floor position, minor repairs such as headlamp damage can be pulled from the end posts.

The fourteen sockets make a complete body bay, and if sited as shown in Figure 112(b) room is available in the central area for a body jig.

The position of the body bay should be selected carefully to ensure that, while the central area is occupied by a major repair, other vehicles outside the posts can be worked on using pairs of sockets. If a body jig is not required, twelve sockets positioned as shown in Figure 112(c) provide adequate flexibility of use. Diagonal pulls are possible between pairs of corner sockets.

Large body repairers often install many sockets throughout a shop. By attaching items of body bay equipment to carefully sited pull posts, several cars with various types of damage can be worked on at the same time, anchorage points being distributed among the remaining pattern of sockets. By installing sockets in this manner, greater flexibility is obtained and waiting time is reduced to a minimum.

The Korek system

The Korek system (Figure 113) is another method of pulling full frame vehicles, unitized bodies and special collision-absorbing structural members. However, additional equipment is required, such as a dedicated bench, in order to accurately repair unitized vehicles that require a more specific measuring system. A disadvantage is that it is stationary and one area of the shop has to be dedicated for frame repairs. However, when not in use there is no problem with using the space for other types of work. Although the base price of the equipment is reasonable, the cost of installation can be quite expensive.

Collision damage repair equipment 93

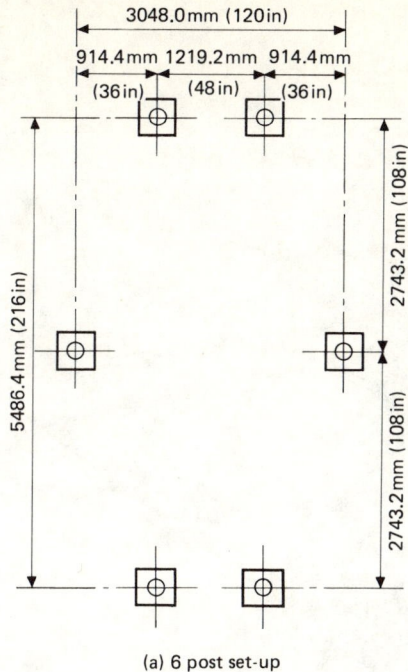

Figure 112 *Layout of pull posts*

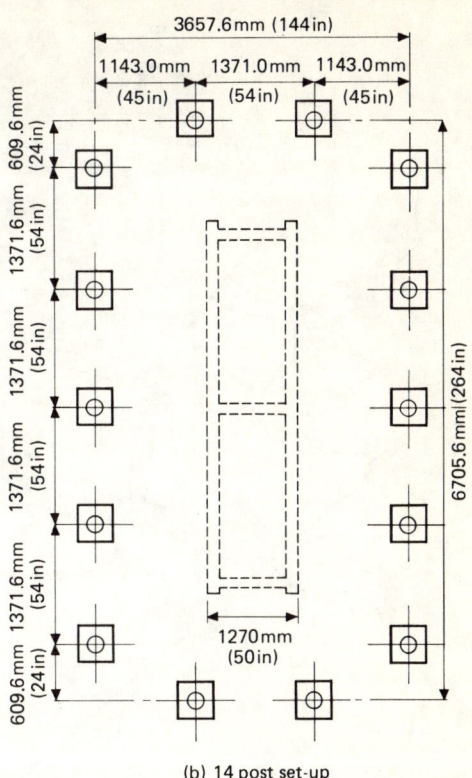

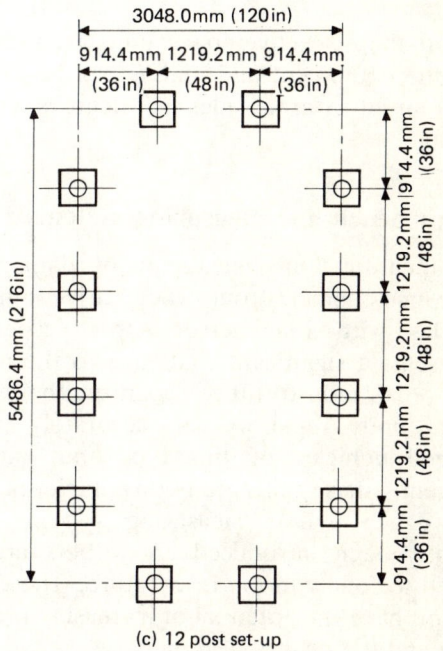

The system operates by placing the car in the centre of the steel frame and using chassis attachments, tubes and safety stands together with chains which are secured to wedged anchoring clamps to retain the car in position.

When pulling is carried out the chains are secured to the vehicle with the aid of pull-type clamps and the opposite end of the chain is wedged securely in the outer frame by means of an anchoring system. The intermediary frame provides the base for the location of a 10 tonne hydraulic thrust unit fitted with a ball head which in turn rests in a recess wedged firmly into the frame.

Pressure is applied on the chains between the two anchoring points. The ball recess arrangements allow the power unit to move freely. The length of the power unit relates to the pulling height and it is easy to adjust since all the extension tubes are fitted with rapid action connectors. Three pneumatic hydraulic foot pumps which can be operated individually to

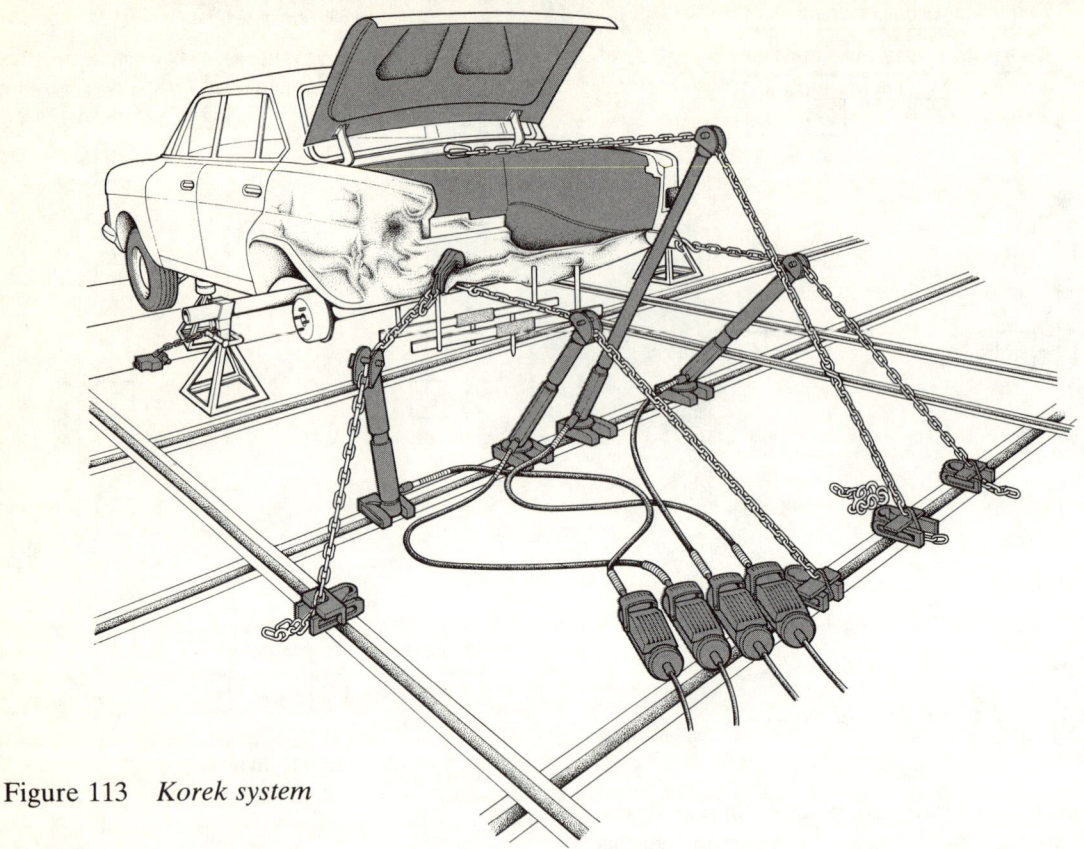

Figure 113 *Korek system*

activate the respective power units are included with the equipment and pulling can be carried out from all external sides 360 degrees around the car.

Repair bench and measuring systems

The identification and repair of damage to transverse-engined front wheel drive unitized vehicles with MacPherson strut suspension represents a significant challenge to the motor body repair industry. It is recognized that highly skilled body repairers can accurately repair unitized vehicles of this type, but all too frequently these skills are not available. Several new bench and/or measuring systems have recently been introduced and others are expected to follow in the near future. These new systems have the potential of increasing productivity and the quality of repair.

Today's need for precision with unitized vehicles

Vehicles currently in production and those scheduled for future introduction will have the features mentioned along with new materials, thinner steels and, in some cases, high-strength steels.

Wheel alignment adjustments are often quite limited or have been totally eliminated. All welded body members are considered to be load carrying. The front and rear portions of the vehicle have been designed with crush zones to maintain passenger compartment integrity. Close tolerances are maintained during production of these vehicles and failure to comply with manufacturers' specifications can result in a multitude of problems such as driveline vibrations, improper ride and road handling difficulties. Through the use of precision measuring methods found on benches/measuring systems, these problems in most cases can be eliminated and damage easily identified.

Examples of vehicle damage which would benefit from a bench/measuring system repair are:

1. Damage to suspension mounting areas.
2. Damage to front rails/aprons that will affect front wheel alignment.
3. Damage to rear rails/inner wheel housing that will affect rear wheel alignment.
4. Damage to the centre section of the vehicle that will affect wheel alignment.
5. Damage to rack and pinion mounting area if mounted to unitized structure.

Methods used to determine damage are:

1. Visual inspection.
2. Diagonal measurements/tram gauge/measuring tape.
3. Frame gauges to establish datum and whether twist or side sway is present.
4. Bench/measuring system.

Dimensions taken should match manufacturers' dimensions (usually available at a dealer) or frame/body specification books, or, if required, dimensions may be taken from an undamaged vehicle.

Classifications of bench systems

All bench-type systems can be described basically as machines designed to straighten unitized vehicles by anchoring the vehicle on top of a heavy metal ladder-type structure and then employing various pulling devices and a precision measuring system to achieve great accuracy in collision repair. The principal difference between the various makes of equipment consists of the measuring method used.

There are two general classes of bench-type systems:

1. The *universal bench* is one that employs movable built-in gauges and laser beams for measuring the accuracy of straightening. The advantage of this type is that any make, year or model of automobile can be straightened with the use of a minimum of equipment. The initial cost as well as the ongoing use cost is lower. However, it relies heavily on the expertise and know-how of the system operator. This is an important variable since a margin for human error must be considered in the use of any bench system.
2. The *dedicated bench* is one that employs jigs or fixtures or templates premoulded, each for a specific make and model of motor car. Therefore, this type of bench will only straighten those cars whose jig or fixture sets are available to the bench operator. Each set of fixtures is made so that it duplicates the car makers' specifications and tolerances. They fasten to key points on the bottom of the car.

A body shop repairing countless makes and models will need many sets of fixtures. This can be costly not only in the initial investment but in the continuing operation as each year new makes and models are introduced. On the other hand there is no room for error in the use of jigs unless of course the wrong fixture or a defective one is used. These fixtures must be either purchased or rented. However, exceptions occur when the floor pan is shared by other makes and models.

Factory specifications are achieved when the control hole and fixture holes line up. Overall this is a designated system for a designated car with the element of guesswork eliminated as much as possible.

Bench-type systems vary in size, weight and price. The larger and heavier machines are usually stationary in that they are bolted permanently to the floor of the bodyshop. An example is the 12 tonne Chief E-Z Liner or the 6 tonne Guy Chart Flex-O-Liner. Lighter machines, such as the Blackhawk Bench, Sun/Celette or the Car-O-Liner, each weighing 0.5 to 0.75 tonne, are mounted on wheels and have the portability feature in their favour. Some are operable with floor posts, others without. Still others are interchangeable.

96 Principles and Practice of Vehicle Body Repair

However, all require that the vehicle be placed on the machine before any straightening can be undertaken.

Examples of bench/measuring systems

Applied Power bench
This equipment (Figure 114) consists of the bench, transverse beams, three sets of fixtures, data sheets and rocker-panel clamp assemblies, and pulling by floor posts or a rail system which allows multiple and 360 degree pulls. The bench is equipped with castors allowing easy movement within the shop. When not in use the bench, which occupies a space of 4191 mm × 1245 mm, must be stored. The working height of the bench is not adjustable and frequently limits access to the vehicle underbody. There are no provisions to compensate for an uneven floor which could cause a twist in the bench and affect fixture fit.

Vehicles are anchored by adjustable rocker-panel clamps and fixtures. The fixtures will also lock in restored areas to prevent distortion. Many of the fixtures have sliding pins that can be lowered for clearance while pulling. The fixtures identify the angle and the position of all critical control points and function as a welding jig to position structural components. Some fixtures are located in areas that require removal of mechanical components.

Using a floor jack one body repairer and one labourer are required for set-up and take-down, which is approximately 2 hours 10 minutes. Using a two-post hoist the time can be reduced to 1 hour 30 minutes.

Car-O-Liner
This equipment (Figure 115) consists of the bench, measuring system, data sheets, rocker-panel clamp assemblies, four rail tie-down legs, pulling units and two jigging clamps. The bench

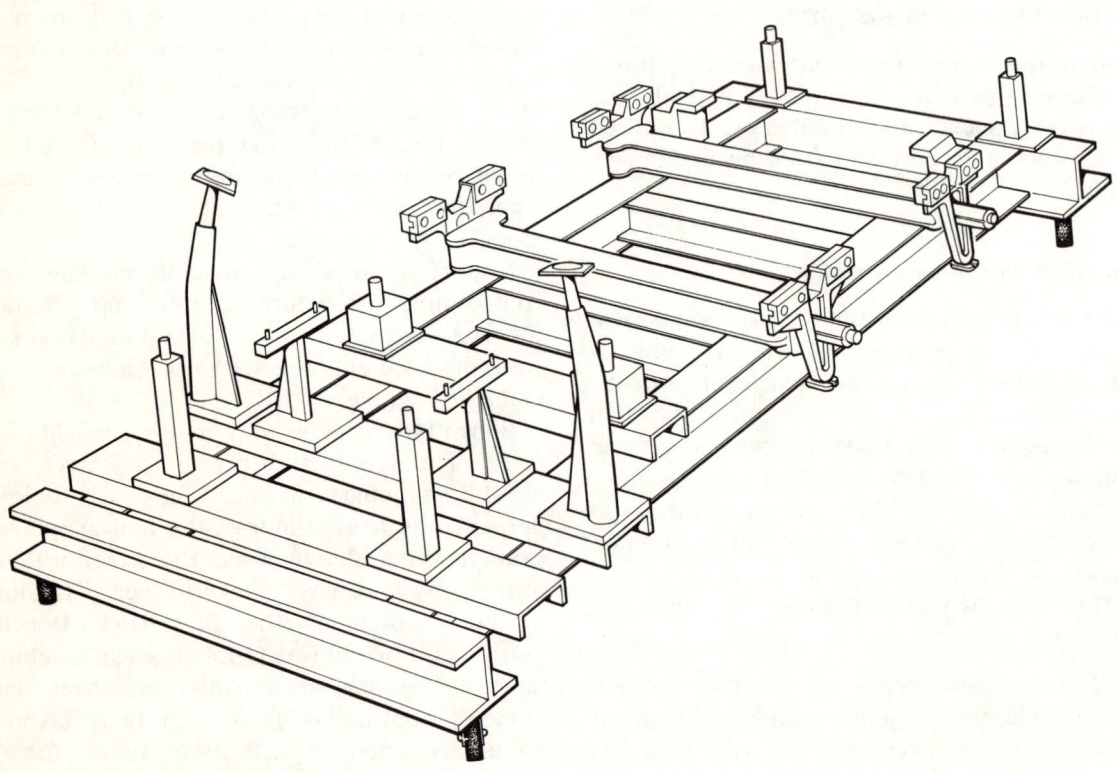

Figure 114 *Applied Power bench*

Collision damage repair equipment 97

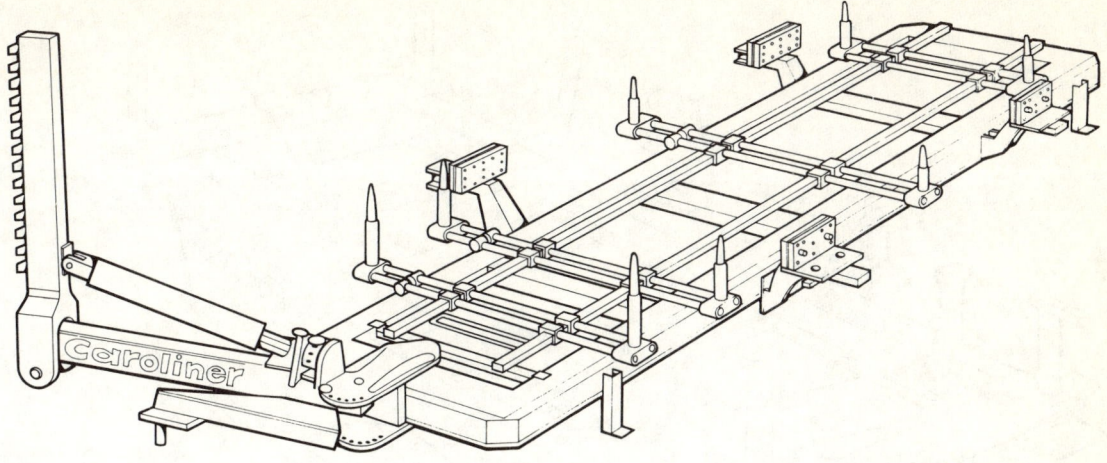

Figure 115 *Car-O-Liner*

is equipped with castors allowing easy movement within the shop and when not in use the bench, which measures 4064 mm × 915 mm, can be set flat on the floor and a vehicle can be parked over it. The working height of the bench is not adjustable and limits access to the vehicle underbody. Sturdiness of the bench prevents any twist from occurring because of an uneven floor. Vehicles are anchored by adjustable rocker-panel clamps and measuring is accomplished by mechanical indicators which identify the proper location, but not the angle, of the control points. There is no provision for locking restored areas; however, jigging for the installation of new parts is provided by special clamps that attach to the measuring indicators and replacement part. Removal of mechanical parts is not necessary with this measuring system and no additional equipment is required to measure different makes and models. Pulling is by using two 'draw-aligners' and also floor posts and rails providing multiple and 360 degree pulls. During pulling operations, the mechanical measuring indicators in the area being repaired should be lowered or removed to prevent them being damaged. After the pull the indicators should be repositioned and checked for proper fit. Using a floor jack one body repairer is required for set-up and take-down, which takes approximately 1 hour 45 minutes.

Sun/Celette bench
This equipment (Figure 116) consists of the bench, levelling legs and gauges, transverse beams, two sets of fixtures, data sheets, rocker-panel clamp assemblies and a pulling unit. The bench is equipped with castors for mobility around the shop. When not in use the bench, which occupies a space of 3988 mm × 1270 mm, must be stored. The working height is not adjustable but levelling legs and a gauge compensate for uneven floors.

Vehicles are anchored by adjustable rocker-panel clamps and the fixtures which will also lock in restored areas to prevent distortion. The fixtures identify the angle and the position of all critical control points and function as a welding jig to position structural components. Some fixtures are located in areas that require removal of mechanical components. The pulling unit is adjustable and will provide a multitude of positions to satisfy the pulling angle required. The pulling unit will allow 360 degree pulls but multiple pulls require the addition of another

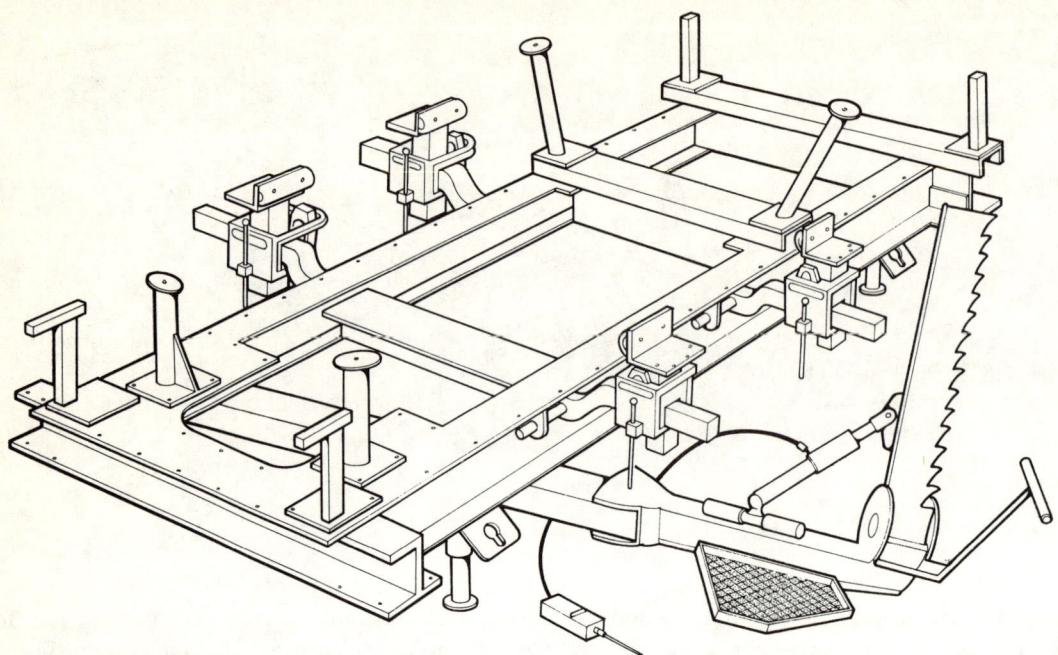

Figure 116 *Sun/Celette bench*

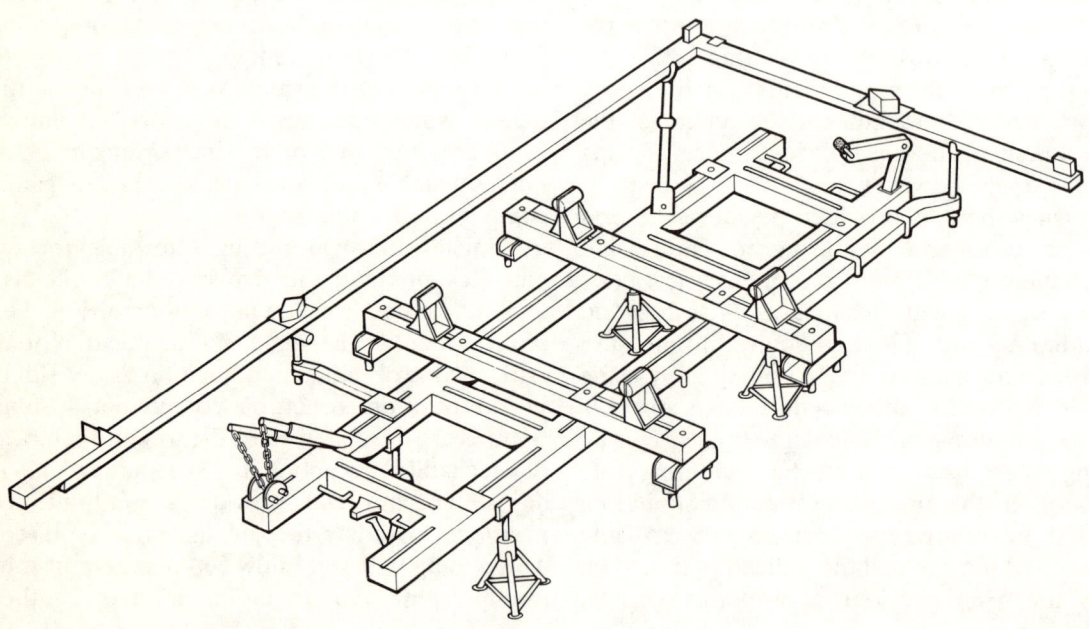

Figure 117 *Nike Dataliner bench*

unit. Pulling can also be accomplished by floor posts and the rails which also allow multiple and 360 degree pulls. Using a two-post hoist, one body repairer and one labourer are required for set-up and take-down, which takes approximately 1 hour 30 minutes.

Nike Dataliner
This equipment (Figure 117) consists of the bench, the laser measuring system, data sheets, transport beams, two pulling beams with trolleys, two pulling units, rocker-panel clamp assemblies, two 'Flex-Loc' arms and four adjustable jack stands. The bench, which measures 4115 mm × 915 mm, is equipped with castors and when not in use can be set flat on the floor and a vehicle can be parked over it. The working height of the bench can be adjusted by the jack stands from 356 mm to 384 mm. This allows convenient access to the vehicle underbody and provides a comfortable working height. The adjustable jack stands also compensate for an uneven floor. Vehicles are anchored by adjustable rocker-panel clamps and measuring is accomplished by the laser beam and plastic scales which will identify the proper location but not the angle of the control point. There is no provision for locking restored areas; however, jigging for the installation of structural components is provided by the two 'Flex-Loc' arms. Removal of mechanical parts is not necessary with this measuring system and no additional equipment is required to measure different makes and models of cars. Pulling is provided by the two pulling beams equipped with the pulling units allowing multiple and 360 degree pulls. After a pull has been made the position of the vehicle should be checked. If its position has not been maintained it will be necessary to recalibrate the measuring system. Using a floor jack one body repairer is required for set-up and take-down, which takes approximately 2 hours.

The Paulee bench – universal and dedicated system
This equipment (Figure 118) consists of a hoist-mounted bench, ramps, a loading trolley, a two-dimensional puller, a three-dimensional puller, two sets of fixtures with data sheets, a universal measuring system with data sheets, a

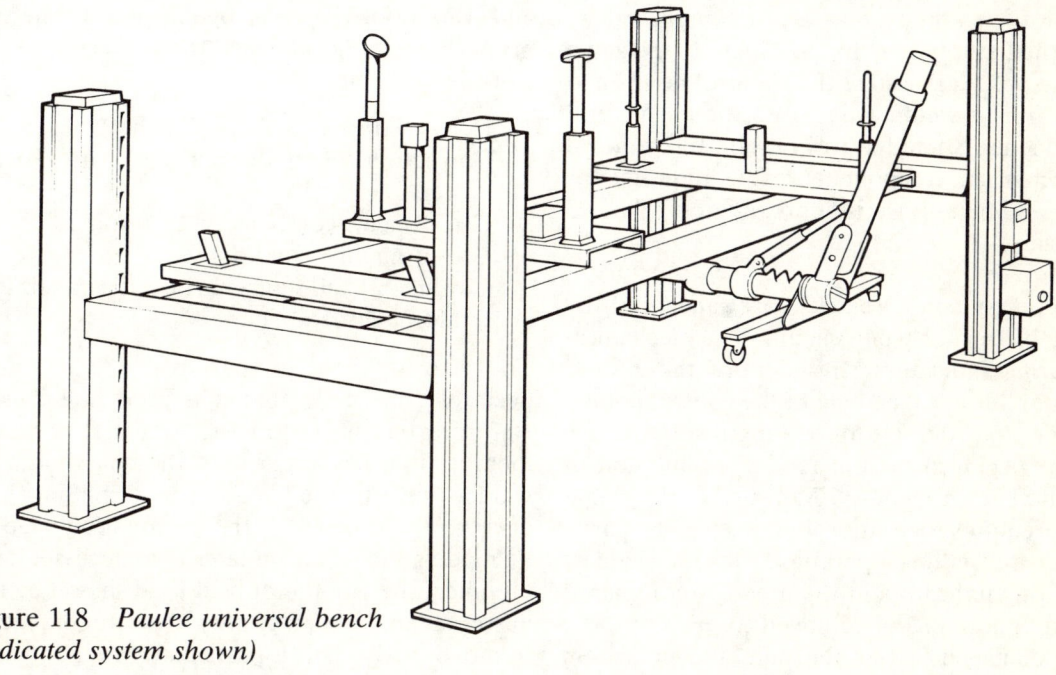

Figure 118 *Paulee universal bench (dedicated system shown)*

mechanical measuring gauge, rocker-panel clamp assemblies and transverse beams.

The hoist-mounted bench is electrically operated and levelled by adjusting cables at each post which are bolted to the floor. The bench occupies a space of 5232 mm × 2819 mm and when not used for structural repairs the hoist can be used for mechanical, other body repairs, or raised so that a vehicle can be positioned under it. The working height of the bench varies from ground level to 1536 mm, allowing convenient access to the vehicle underbody and providing a variable working height. The hoist-mounted bench is not adaptable to floor posts and rail systems currently used in body shops. The pulling equipment provided allows multiple and 360 degree pulls at any bench height by use of a two- or three-dimensional puller using either the dedicated or the universal system. Two body repairers are needed to mount the vehicle on the loading trolley. Once on the trolley one body repairer is required for set-up and take-down.

Dedicated system This is a positive system that eliminates the need for additional holding by rocker-panel clamps. Vehicle anchoring and locking of restored areas are accomplished by using the fixtures. The fixtures identify the angle and the position of all critical control points and function as a welding jig to position structural components. Some fixtures require the removal of mechanical components. Time required for set-up and take-down is approximately 1 hour 5 minutes.

Universal system Vehicles are anchored by adjustable rocker-panel clamps. The mechanical measuring indicators will identify the proper location, but not the angle of the control points. There is no provision for locking restored areas and no jigging provision for the installation of structural components. Measuring, in some areas, requires the removal of mechanical parts. During the pulling operations, the mechanical measuring indicators in the area being repaired should be lowered or removed to prevent them being damaged. After the pull, the indicators should be repositioned and checked for proper fit. If the vehicle position has not been maintained it will be necessary to reposition the vehicle to the measuring system. A mechanical measuring gauge is also available which will measure MacPherson strut tower locations without mechanical part removal as well as upper body locations. Time required for set-up and take-down is approximately 2 hours 25 minutes.

General repair techniques

In general the method of repair is to analyse the crash, establish the order in which damage occurred, and reverse the order when correcting the damage. Obviously, distortion to the frame or chassis must be rectified first. If bodywork is pulled or pushed back into proper line without first correcting frame damage, the body will never hold its alignment. Also, in many cases where, for instance, a door is itself undamaged but does not line up properly with the door opening, individual realignment is not necessary. When the frame or chassis is restored to true, the undamaged bodywork will come with it and the door and its opening will line up properly once more. Thus, the correct order of work is:

1 Heavy external pulling with the Dozer.
2 Correction of body damage with Porto-Power.
3 Use of heat, combined with 1 and 2.
4 Use of hammer and dolly.
5 Cutting out and replacing buckled panels and reinforcing members.

The use of heat is permissible and often desirable, provided that it is used judiciously. Before a pull is started on any panel or frame member that has a tear in it, the tear should be welded up. If this is not done pulling will aggravate the tear and the eventual damage to be made good will be greater than necessary. At all times, the amount of heat used should be the absolute minimum necessary to permit the metal to move freely. This is particularly so in the case of frame members where part of the strength of the member derives from internal stresses set up during the forming operations which are part of the manufacturing process. Heat helps to relieve these stresses and excessive heat can impair the load-carrying capabilities of the member. The fact that heat helps to relieve stresses is of considerable importance in correcting damage to sections that have been creased or buckled as a result of the impact. Here, the stresses resulting from the buckles help to retain the metal in its damaged state; even after it has been restored to its correct contour by the application of pressure, when that pressure is removed the section will return to its damaged position. Therefore, before the pressure is taken off, these stresses must be relieved by ironing out the buckles or creases with heat used in conjunction with the hammer and dolly.

The cutting out and replacement of damaged panels and reinforcing sections should always be one of the last jobs tackled. Often, it will be advisable at least to rough out the damage in a section that is eventually going to be replaced. This may seem at first sight to be a complete waste of effort, but it should be remembered that the successful fitting of a replacement section will depend on the correct alignment of the surrounding areas, and this can best be established by restoring roughly the true shape of the badly damaged section before it is removed.

6 Refinishing facility and equipment

The refinish facility

Refinish departments are developed according to the type of vehicles, volume of refinish work and paint systems used. In all developments a primary requirement is accessibility – for both customers and operators. A clean, well-equipped and properly organized refinish area is essential for consistent high-quality work at maximum throughput. Sequential flow of vehicles through the premises is necessary, with a record system for the larger establishment, and must be given major consideration when planning new premises or extending an existing facility.

Layout and equipment must be planned to cover current requirements and should allow for possible future development, with provision for all aspects of health, safety and training regulations. Labour materials and space are best utilized by providing operators with the right equipment and facilities to ensure proper control of paint temperature and viscosity and planned economical movement of vehicles.

Simplicity of vehicle movement assists in the continuous use of spray booths and fosters maximum efficiency to meet the varied requirements demanded of the refinisher. Where the working area is small or restricted, the installation of a turntable system will in many cases improve vehicle movement and accelerate throughput. The methods of moving vehicles within the refinish shop are:

1 *Under their own power* This method is the most economical in terms of manpower, but requires the maximum space. Also it gives rise to additional dirt, air pollution and fire hazards.
2 *Manually* Again uses maximum space and the demand on manpower is heavier.
3 *By mobile hydraulic jacks* By allowing a tight turning circle both front and back, these permit some saving of space.
4 *By turntable* Use of a turntable (Figure 119) offers an excellent means of utilizing what could otherwise have been a dead corner. Or it may permit the spray booth to be sited conveniently in what would otherwise have been an impossible position.
5 *By the rail and bogie system* In this, the vehicle is moved along tracks on bogies placed under each of its four wheels (Figure 120). This offers the greatest advantage of

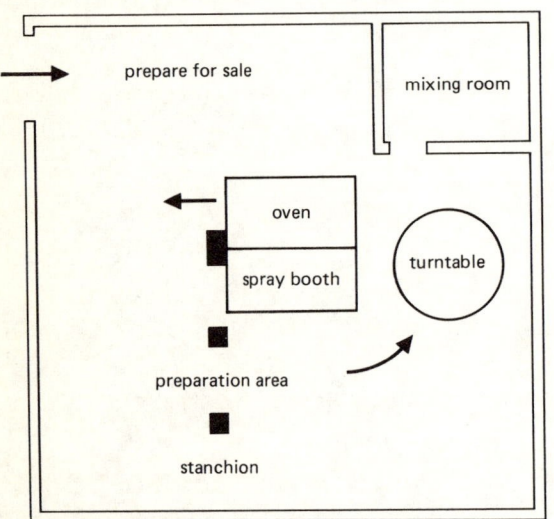

Figure 119 *Paint shop layout showing use of turntable*

Refinishing facility and equipment 103

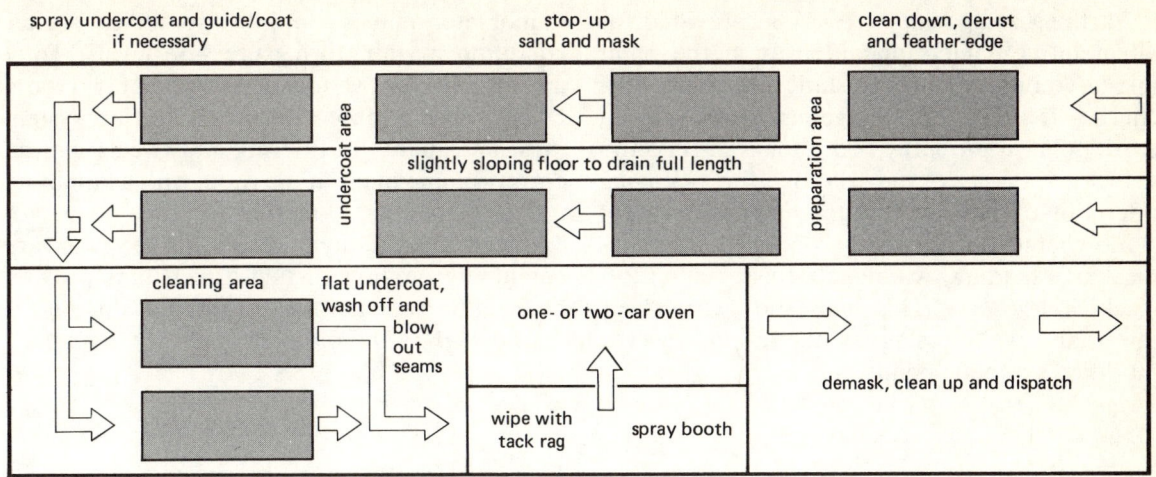

Figure 120 *Paint shop layout with lateral movement on rails*

all-sideways movement of the vehicle. Many different layouts are possible. For example, four sets of parallel rails could be used. The first pair constitutes the track on which the vehicles are first received into the paint shop, and along which the preparatory work is usually carried out, such as discing, flatting and stopping. The second pair is kept clear to allow the movement of selected bodies to the spraying and/or stoving positions. Though movement is more restricted than with two tracks, it is also possible to have layouts utilizing only one track.

In all establishments, mechanical and body operations should ideally be divorced from the paint area, within which further separation is considered essential to segregate preparation bays with drainage for wet flatting, steam cleaning, etc. from the spraying area. Two vehicle parking spaces should be provided for each preparation bay, and an additional separated area with a lift for steam cleaning, washing and underbody finishing is desirable. A compressed air supply and normal mains services, electric, water, drainage, and gas (if a gas-fired heating system is to be employed) are essential.

Water supply and drainage

Refinish premises must be supplied with water, both for personnel and vehicle cleaning facilities, for wet flatting and water-washed dust extraction systems. Hoses fitted with an adjustable pressure nozzle or a high-pressure water wash or steam cleaning unit should be available for removing general surface and underbody dirt accumulation. Preparation bays should be supplied with a suitable water outlet to permit hosing and washing of floors and vehicles as necessary.

Electrical

All electrical appliances and fittings must comply with the relevant standards appropriate to their use and environment. Compliance with the local regulations regarding flame-proof installations in storage, mixing and spraying areas is necessary. Use of conduit fittings, trunking and encapsulation are implicit requirements throughout the refinish premises. Electrical sockets/plugs, switches, outlets, etc. must be positioned on walls at a height which will ensure no ingress of water but is within easy reach of operators.

Earthing arrangements must be provided for all metal containers and utensils in the paint mixing room to prevent static electricity discharge. It is also advantageous to provide an earth point in the spray booth for connection to vehicles prior to and during the spraying operation to reduce dust attraction as a result of static charges from rubbing operations such as masking, sanding, wiping, flatting, tacking off (tack-cloth), etc. This may be particularly important when using synthetic and acrylic modified synthetic paints.

Heating, ventilation and dust extraction

The building must be free from draughts, properly heated, ventilated and provided with sufficient means of extraction to protect employees from airborne contaminants. The siting of such equipment is conducive to the efficiency of the total refinish operation, particularly the heating and ventilation units. Air should be extracted at a low level to enable a roof to floor circulation system. The heating facility should be capable of maintaining an ambient working temperature at approximately 20 °C.

Owing to the air extraction requirements for spray and dust removal from the atmosphere, it is prudent to install filtration, extraction and heating as a composite recirculatory ducted unit to comply with relevant regulations, or to incorporate separated areas where extraction is required for spraying primer, sanding, etc. Single or multiple dust arrestor units can be used in open preparation areas and partitioned primer spraying areas. The capacity of the air extraction system required in a partitioned refinish area will be less than that required for an open area, and this will also be reflected in greater efficiency and lower operating costs, as the units can be used according to the needs of the operators.

The filter elements must be capable of easy removal or exposure of fresh surfaces by peeling off the skin layer when the filtration becomes impaired with dust or dry paint. A water manometer may be utilized to indicate filter condition by monitoring pressure levels. Total air filtration flow should be at a rate of two room changes per minute and the extraction system must be capable of filtering and circulating the entire volume of air in the room with a minimum air velocity of 30 m/min at the points of extraction. Although direct drive fans are commonly employed, belt-driven fans provide high efficiency as the extraction ducting is unimpeded by the motor.

Lighting, decoration and headroom

Clean conditions and good light are essential for obtaining good-quality work. Lighting to daylight standard must be installed throughout the refinish area, with smooth surfaced walls, ceiling and masonry, rendered dust free with appropriate white or light grey paint finish. The roof should have a minimum head clearance of 3 m (10 ft).

Floors

Floors must be sealed or painted a light colour, preferably grey, providing a smooth surface to discourage dust and dirt and enable easy cleaning. Adequate surface contour and drainage, or floor grilles etc. to enable frequent hosing are essential to reduce the risk of dirt contamination.

Administration

An administration area and defined reception area must be sited in a prominent position for customer recognition and adjacent to the refinish department. Progress of operator time and work within large refinish establishments can be conveniently controlled by utilizing a visual record chart system.

Paint store

A separate paint store or storeroom is essential for all establishments and may be either a portable unit or a permanent structure. In smaller establishments a secure approved cabinet or vault may suffice. If storage is detached from the main building, it should be fully heat insulated to prevent paint from approaching very low temperatures during cold weather as cold paint necessitates excessive thinning. Paint at the correct working temperature will always take the correct amount of thinner.

Compressed air supply and equipment

Compressor

The compressor (Figure 121) may be a screw, vane or reciprocating piston type. The piston type is the most commonly used in the refinish trade, although very large and continuous air demands can be economically met with the modern screw and rotary types with reduced noise levels.

The compressor must be of sufficient capacity to cater for all operator and work loads. The output rating in cubic metres, litres or equivalent cubic feet, per minute, must be calculated on the basis of actual total air consumption when all equipment is in use. Compressors may be rated according to piston displacement, displacement at a specified pressure, or free-air displacement at various speeds. The rating which equates to actual output is the free-air displacement at operating speed as this will depend on the volumetric efficiency, which is approximately 75 per cent of the piston displacement for single-stage compression and 85 per cent for two-stage.

For pressures under 5.95 bar (85 psi), single-

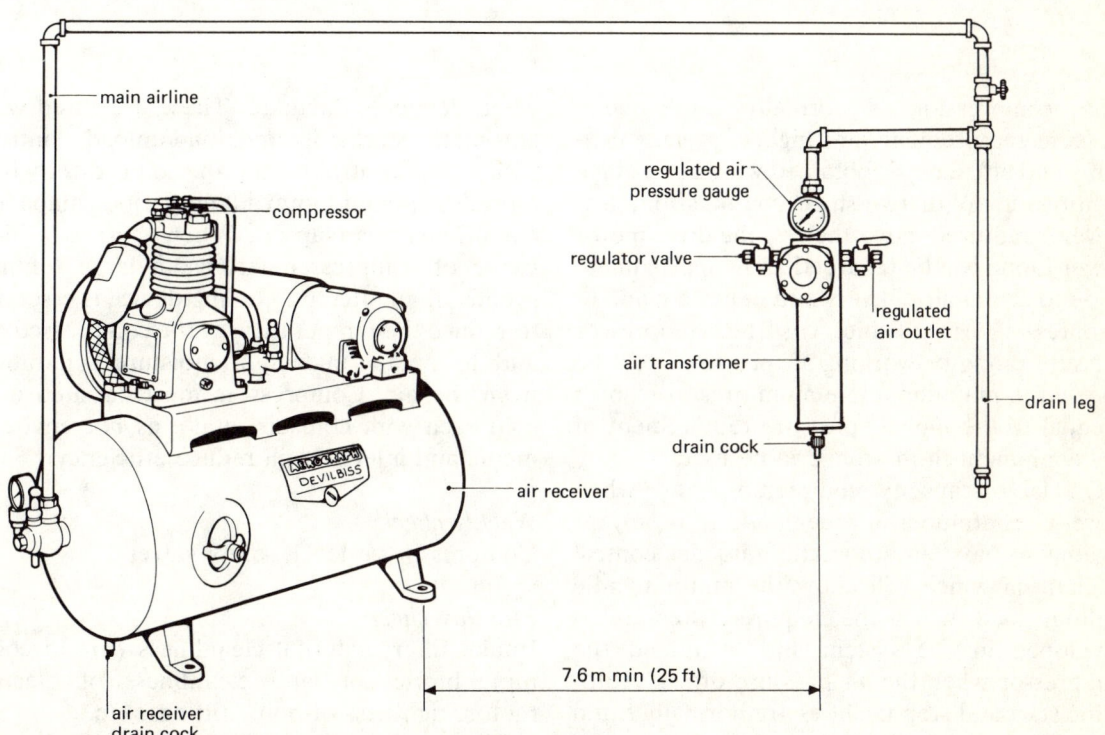

Figure 121 *Air compressor and transformer/regulator*

Table 4 Guide to compressor size

Average pressure (bar)	(psi)	Continuous air consumption (m^3/min)	(ft^3/min)	Intermittent air consumption (m^3/min)	(ft^3/min)	Required motor power rating (kW)	(Hp)
7	(100)	0.11–0.16	(4–6)	0.40–0.60	(14–22)	1.20	(1.50)
10.5	(150)	0.08–0.14	(3–5)	0.34–0.51	(12–18)	1.20	(1.50)
7	(100)	0.16–0.23	(6–8)	0.62–0.85	(22–30)	1.86	(2.50)
10.5	(150)	0.14–0.19	(5–7)	0.51–0.68	(18–24)	1.86	(2.50)
7	(100)	0.23–0.37	(8–13)	0.85–1.30	(30–46)	3.00	(4.00)
10.5	(150)	0.19–0.28	(7–10)	0.68–1.02	(24–36)	3.00	(4.00)
7	(100)	0.37–0.57	(13–20)	1.30–1.70	(46–60)	4.10	(5.50)
10.5	(150)	0.28–0.48	(10–17)	1.02–1.44	(36–51)	4.10	(5.50)
7	(100)	0.57–0.82	(20–29)	1.70–2.07	(60–73)	5.60	(7.50)
10.5	(150)	0.48–0.74	(17–26)	1.44–1.87	(51–66)	5.60	(7.50)
7	(100)	0.82–1.13	(29–40)	2.07–2.83	(73–100)	7.45	(10.00)
10.5	(150)	0.74–0.99	(26–35)	1.87–2.50	(66–88)	7.45	(10.00)
7	(100)	1.13–1.67	(40–60)	2.83–3.60	(100–127)	11.20	(15.00)
10.5	(150)	0.99–1.50	(35–53)	2.50–3.74	(88–132)	11.20	(15.00)
7	(100)	1.67–2.41	(60–85)	2.83–4.36	(127–154)	18.60	(25.00)
10.5	(150)	1.50–2.27	(53–80)	3.74–5.66	(132–200)	18.60	(25.00)

stage compression is normally used; when pressure requirements are higher, greater economy and efficiency is obtained with a two-stage compressor. With two-stage compression, less power is required; nevertheless, the drive motor power rating can be regarded as an approximate guide to actual air delivery and hence a guide to compressor size (Table 4). The compressor pressure rating or working air pressure must be selected to maintain a minimum pressure which is equal to the highest pressure requirement of any equipment in the range to be used.

On larger-capacity compressors and when there is continuous air demand, it is advantageous to have an automatic unloader control mechanism which will allow the motor to idle without load when the required pressure is developed in the system and to reload the compressor when the air pressure drops. Automatic start and stop facilities are normally fitted to small compressors for which the air demand is intermittent. Dual-control versions may be used when demands fluctuate. These are fitted with automatic start/stop and load/unload controls with a time control to stop the compressor when a predetermined motor idling period (unloaded conditions) has elapsed. When there is a high usage of compressed air, as in large refinish premises, an effective cooling device is essential to reduce the temperature of the compressed air and to obtain maximum moisture separation from the air. Compressors must be sited in a cold area with cold air intake as hot environments and inlet air will reduce efficiency.

Weekly check
Compressor oil level, safety valves.

Monthly check
Intake filter, external cleanliness (to aid cooling), lubrication and cleanliness of electric motor, tightness of bolts, fittings, etc.

The compressor reservoir must be of sufficient capacity to eliminate air pressure fluctuations from the output of the compressor and to sustain short periods of peak demand. The reservoir also serves as a water separator and must be drained daily.

Main airlines

The main airlines (see Figure 121) must be free from air leaks and manufactured from galvanized iron. A single-feed system with an internal bore of 38 mm (1.5 in) will prove adequate for air consumption volumes up to 2.8 m^3/min (100 ft^3/min) up to pressures of 10.5 bar (150 psi) and total pipe length of 76 m (250 ft) without excessive air pressure drop, and will also permit future expansion and development of facilities or the installation of a larger compressor.

A smaller-bore main airline can be used if the system is designed as an endless loop, when a 25 mm (1 in) internal bore line can be used as an equivalent to the single-line system with 38 mm (1.5 in) bore. Main airlines must have a water accumulation and drain point and reduced bore sizes only where necessary to accommodate the fittings at the various outlet points. The compressed air supply should be arranged to provide each operator with two outlets assuming one operator per vehicle work area. If the main airline is more than 10 m (30 ft) from compressor to the initial outlet a main airline oil and water filter must be incorporated close to the compressor to prevent line contamination.

Air tool oilers

An in-line lubrication unit (oil drip) may be fitted to outlets which will be used exclusively for pneumatic power sanders and tools. Lubricators must not be incorporated where airlines serve spray guns or blow guns. Alternatively, small-capacity line lubricators are available for fitting directly onto the air tool inlet and which obviate the need for main line lubricators.

Air regulators/filters (transformers)

Air regulators (see Figure 121) must be of sufficient air flow capacity for the largest demand at the output and will be rated accordingly and fitted only where a regulated supply is required at spraying points. Filters (or separators) must also be of the correct air flow rating and are normally integral with the transformer. A filter with drain cock must be incorporated after the regulator and must not be allowed to become clogged with oil or filled with water which condenses from the air passing through it. Frequent draining is essential in conjunction with periodic cleaning or renewal of the filter as appropriate. Failure to maintain filters will result in refinish defects. To be efficient, transformers must not be installed adjacent to the compressor, or near heating appliances. Each final output from the regulator/filter unit should be fitted with a shut-off valve prior to the female hose coupling. Small-capacity moisture filters are available for fitting directly onto the spray gun inlet tail (between the air hose coupling and the gun) but these must *never* be regarded as or used as a substitute for main airline filter units.

Air hoses

Hose material should be selected for flexibility, chemical resistance and pressure rating. The use of excessive length airlines is to be deplored as this causes pressure drop (see Table 5) and can cause mottle, sags and slow drying, the result of poor atomization. Hose lengths of 6.1–9.1 m (20–30 ft) with a minimum internal diameter of 7.9 mm (5/16 in) should be used. Spray points should be planned so that this length of hose need never be exceeded.

Quick-release couplings

These allow a rapid interchange of air-powered equipment, blow guns and spray guns. A male

Table 5 Air pressure drop per hose length

Pressure at regulator bar (psi)	Length of hose m (ft)						
	1.5 (5)	3.0 (10)	4.6 (15)	6.1 (20)	9.1 (30)	12.2 (40)	15.2 (50)
	Working pressure at the spray gun bar (psi)						
6.3 mm (¼ in) bore							
2.1 (30)	1.8 (26)	1.7 (25)	1.7 (24)	1.6 (23)	1.5 (21)	1.3 (18)	1.0 (14)
2.8 (40)	2.4 (34)	2.2 (32)	2.2 (31)	2.0 (29)	1.7 (25)	1.5 (21)	1.2 (19)
3.5 (50)	3.1 (44)	2.9 (41)	2.7 (38)	2.5 (36)	2.2 (32)	2.0 (28)	1.6 (23)
4.2 (60)	3.6 (52)	3.4 (48)	3.2 (45)	3.0 (43)	2.7 (39)	2.4 (34)	2.1 (30)
4.9 (70)	4.2 (60)	4.0 (57)	3.8 (54)	3.6 (51)	3.2 (46)	2.9 (41)	2.6 (37)
5.6 (80)	4.8 (68)	4.5 (64)	4.3 (61)	4.0 (57)	3.6 (52)	3.4 (48)	3.1 (44)
6.3 (90)	5.4 (77)	5.1 (73)	4.8 (69)	4.6 (65)	4.1 (59)	3.9 (55)	3.5 (50)
7.9 mm (5/16 in) bore							
2.1 (30)	2.0 (29)	2.0 (28)	1.9 (27)	1.8 (26)	1.8 (25)	1.7 (24)	1.6 (23)
2.8 (40)	2.7 (38)	2.6 (37)	2.6 (36)	2.5 (35)	2.4 (34)	2.3 (33)	2.2 (32)
3.5 (50)	3.3 (47)						2.8 (40)
4.2 (60)	3.9 (56)	Subtract from initial pressure drop approximately 0.028 bar per m (0.5 psi per 5 ft)					3.4 (49)
4.9 (70)	4.6 (66)						4.0 (57)
5.6 (80)	5.3 (74)						4.6 (66)
6.3 (90)	5.9 (84)						5.3 (74)

connector is normally fitted to one end of the hose to mate with the female on the main airline filters, and the opposite end has a female connector to accommodate the male fitted on the equipment. This arrangement permits complete detachment of air hoses from equipment and transformers, allows short hoses to be fitted together and reduces the possibility of twisting and tangles as the couplings permit separate rotation of the tool and hose.

Alternatively the hose may be attached directly to the air output side of the regulator/filter when these are fitted with a shut-off valve or tap.

All connectors, tails and fittings must be securely attached to the air hoses with suitable O rings or clips.

Preparation

Two preparation/working bays per operator, each having an area of 6.2 m × 3 m (20 ft × 10 ft) are minimum requirements. The preparation/prime area must have adequate lighting, drainage, ventilation and heating. A separate cleaning and retrimming area will reduce the risk of contamination from the spraying areas, especially when primer spraying areas and preparation bays are also segregated by suspended anti-static, transparent, strip dividers and doors. Anti-static strip dividers are practical and advantageous for bay segregation and for a clean enclosure at the spray booth entrance where the vehicle can be finally blow dried, dusted off, spirit wiped, etc. prior to colour coat application in the booth.

Mixing room

A small separate room, sited adjacent to the spray booth operator access door, with extraction, heating and all electrical fittings in compliance with relevant standards, provides a safe, efficient and clean environment for colour mixing and matching. The colour mixing scheme is considered essential for direct availability of the full range of colours and must be supplied by an approved paint manufacturer. A stainless steel work surface, a cabinet with shelving and gun cleaning equipment should be installed in the mixing room. Ancillaries such as a colour matching light box, a small oven for drying sprayed test panels, a filing system for test colour panel reference, etc. may also be included.

Mixing scheme equipment

Mixing scheme combined shelving and stirring units are available in various configurations and sizes with from 6 to 40 stirring heads to suit the requirements of the refinisher regardless of the type and quantity of work. They are offered with electric or compressed air driven motors driving the stirring heads through belt, chain or mechanical arrangements. Compressed air powered motors are normally used for the smaller mixing schemes having 10–15 stirring heads. The units with 25–40 stirring heads are usually electrically powered. The units should be provided with guards, and with paint tin retaining clips at each stirring position. The electric motor and microfiche unit must be to an approved flame-proof standard. Equipment lease is offered by paint manufacturers, as well as rapid, motorized, individual stirring heads or hand stirring attachments, so that users with low requirements can retain the benefits of colour mixing without high investment costs.

Gun cleaning receptacle

In many refinish shops, large quantities of thinner are wasted when cleaning equipment. Economical use of both solvent and time will be achieved by utilizing brushes and a container or 'sump' of cleaning solvent, fitted with a lid to reduce solvent evaporation.

Spray booths

The spray booth (Figure 122) defines the spraying area and introduces high levels of efficiency, cleanliness and health controls to ensure a safe environment for the operator and those in the peripheral area, and provides the most suitable conditions to obtain high-quality refinishing.

Local authority approval may be required before installing a spray booth; the refinisher must also consult the local Fire Officer and Factory Inspector who may specify details of booth design and location requirements. Failure to construct or install a spray booth designed to the basic principles required will necessitate continual update as legislation demands. Whether purchasing or constructing a spray booth, careful attention to siting, manufacture, type of structure and the operating equipment will minimize fire hazards, installation, heating costs, running costs and the needs for further development, and will permit easy entry and movement of vehicles.

An effective spray booth necessitates several basic requirements:

1. Size approximately $7 \times 4 \times 3$ m ($23 \times 14 \times 9$ ft).
2. Filtered input warmed air. Filters should be of the disposable type, and made of either impregnated fibreglass, impregnated textile or open-celled polyurethane foam. Blown warm air is the most efficient form of heating for the spray booth.
3. Adequate insulation to minimize heat loss (a double-skinned structure is recommended for prefabricated 'combi-units').
4. The air flow should be effectively directed from the ceiling to the extraction grilles or ducts at floor level.

110 *Principles and Practice of Vehicle Body Repair*

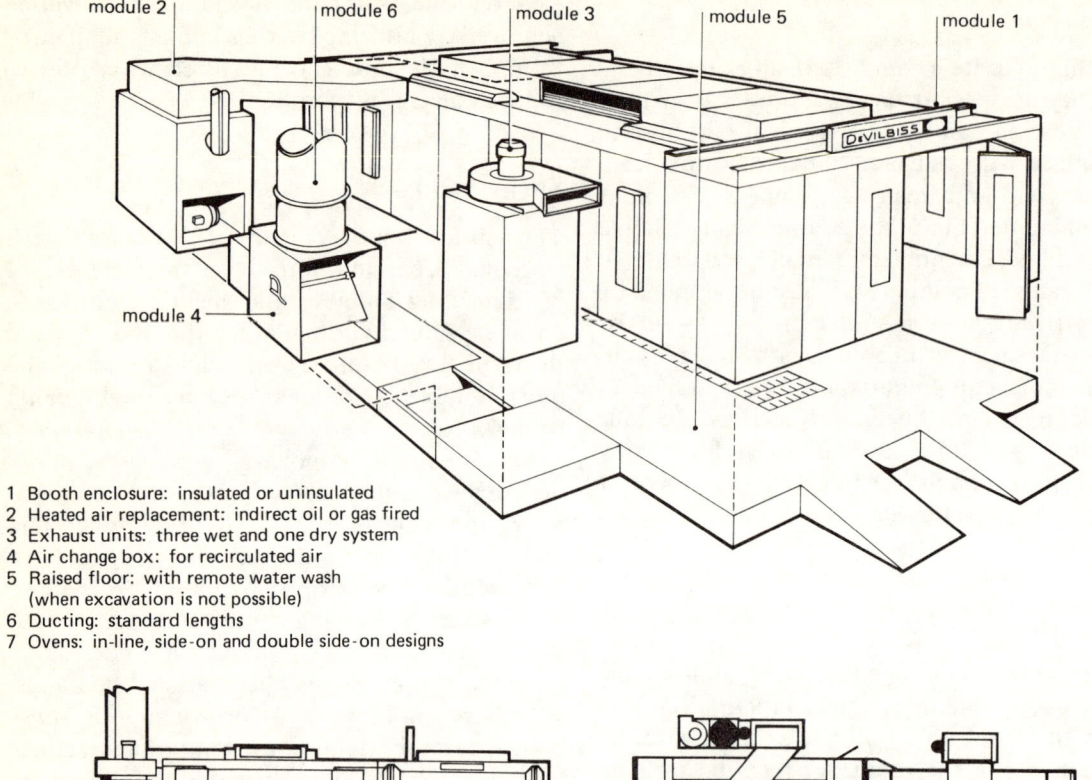

1 Booth enclosure: insulated or uninsulated
2 Heated air replacement: indirect oil or gas fired
3 Exhaust units: three wet and one dry system
4 Air change box: for recirculated air
5 Raised floor: with remote water wash
  (when excavation is not possible)
6 Ducting: standard lengths
7 Ovens: in-line, side-on and double side-on designs

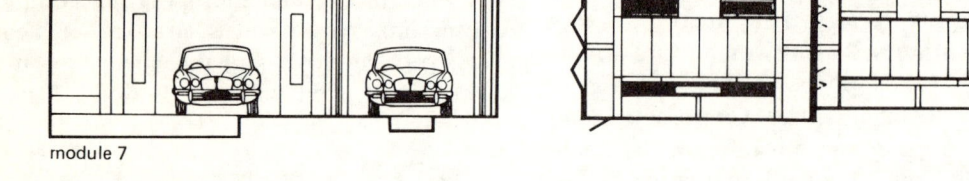

Figure 122 *Modular construction spray booth*

5 When air is to be discharged into an environment which dictates very low levels of atmospheric pollutants, a water-washed filter extraction facility will be the most effective method of reducing the quantity of dry spray, etc. vented to the atmosphere. Such installations also require less ducting maintenance as there will be a lower accumulation of paint deposits in the ducting and on the fans such that their efficiency will be maintained for longer periods. Water-washed filtration may be incorporated as an underfloor arrangement or, more simply and more economically, as an integral part of the extraction system.

6 The circulation system must achieve two complete changes of the total air volume per minute at an approximate input velocity of not more than 40 m/min (120 ft/min).

7 The air extraction must be uniform, i.e. the air flow is evenly extracted all round the vehicle when situated in the booth. Complete air changes ensure efficient ventilation and the elimination of almost all airborne contamination, providing that filters are maintained regularly.

8 Fluorescent lighting situated along the junction of the walls and ceiling, or vertically on the walls, is necessary to provide daylight conditions and must be enclosed with wire-reinforced glass panels, or flame-proof lights. Additional lighting may be required at operator waist height.

In smaller establishments the spray booth may be of a non-heated dry extraction type, providing that the general heating requirements are adequate to maintain the booth air temperature at approximately 20 °C. A fully equipped temperature-controlled spray booth provides heating by electric-, gas- or oil-fired systems through a heat exchanger to give clean warm air at 20–21 °C which is the ideal ambient temperature for both paint application and operator comfort. Temperature control enables work to continue regardless of external conditions. The provision of a thermometer or indicator and thermal control is essential for maximum fuel economy.

The spray booth may be operated as a pressurized system, when the pressure inside will be greater than outside thus reducing the risk of dirt entry (see Figure 123 and Figure 124) or it may be a non-pressurized air circulatory system (see Figure 125 and Figure 126). In an unpressurized booth, it is essential that all doors are effectively sealed to prevent dirt ingress when the extractor fans are running.

Premanufactured spray booth units are commonly installed and these must have proper roof clearance with correct ducting to atmosphere.

Maintenance and fault rectification time must be minimal; filtration units that are easy to remove when impaired by paint are an asset to the refinisher. A water gauge manometer is often incorporated in the design of the booth as an indicator of pressure levels and filter condition.

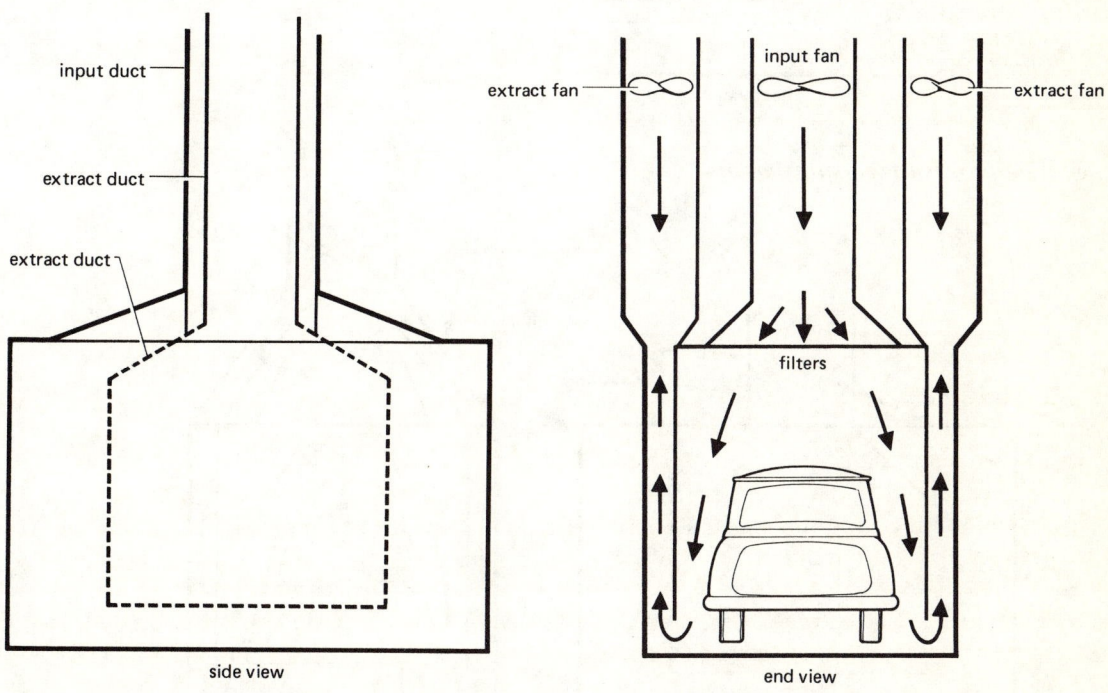

Figure 123 *Pressurized spray booth*

112 *Principles and Practice of Vehicle Body Repair*

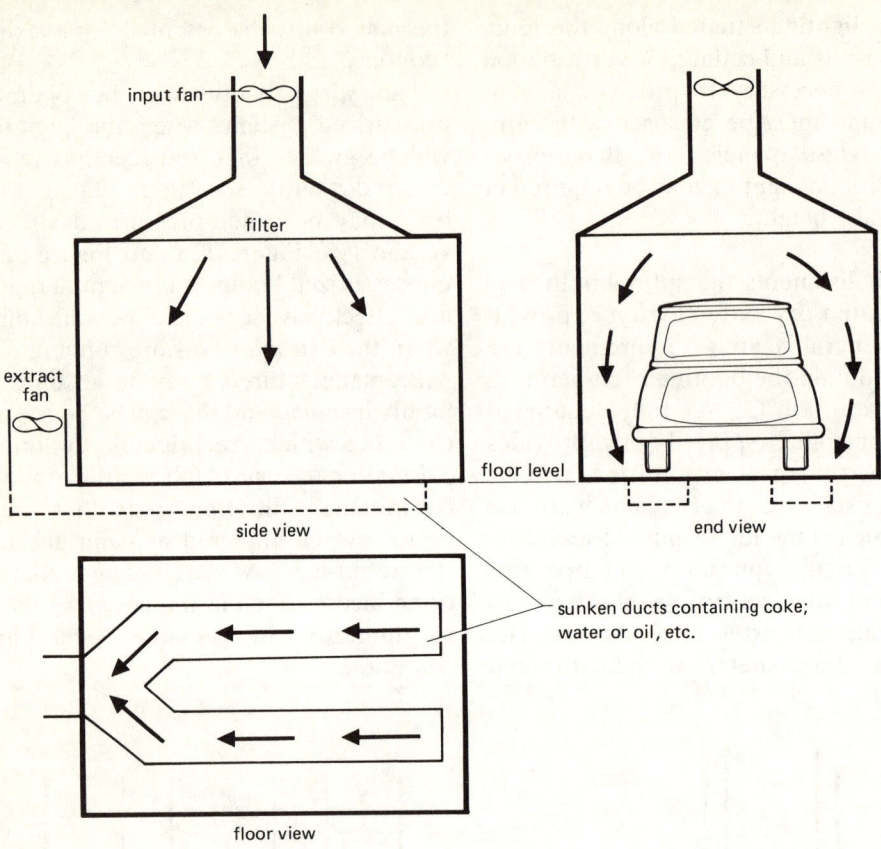

Figure 124 *Pressurized spray booth*

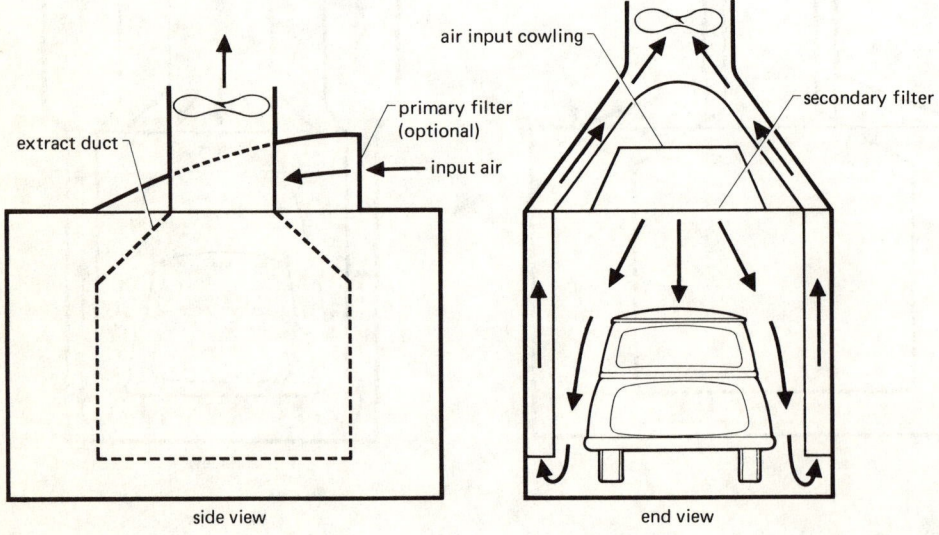

Figure 125 *Simple spray booth (unpressurized)*

Refinishing facility and equipment 113

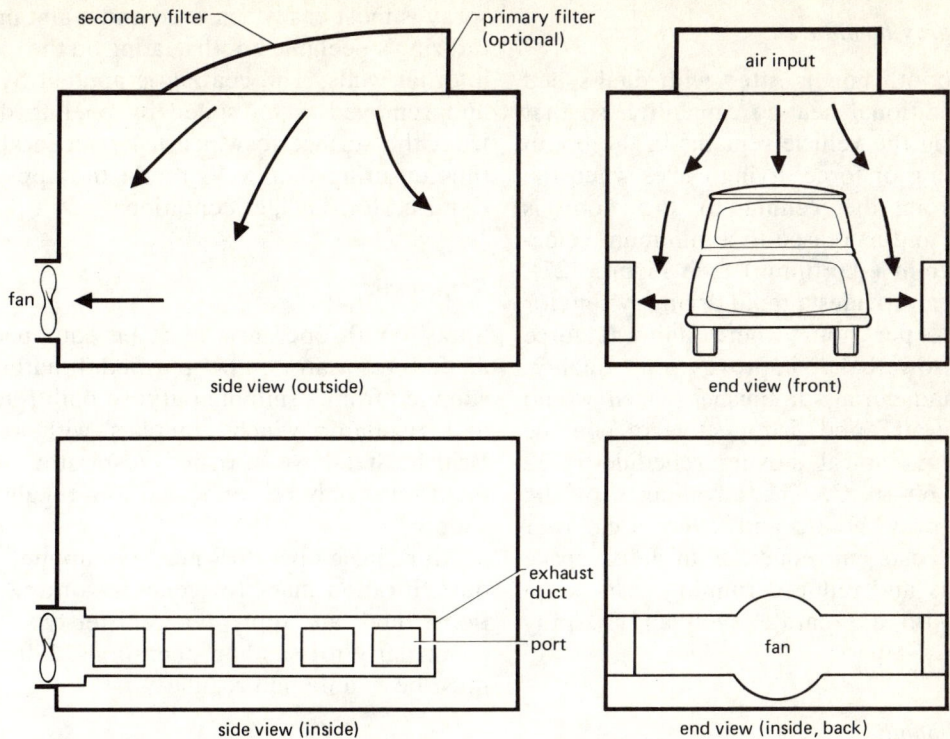

Figure 126 *Simple spray booth (unpressurized)*

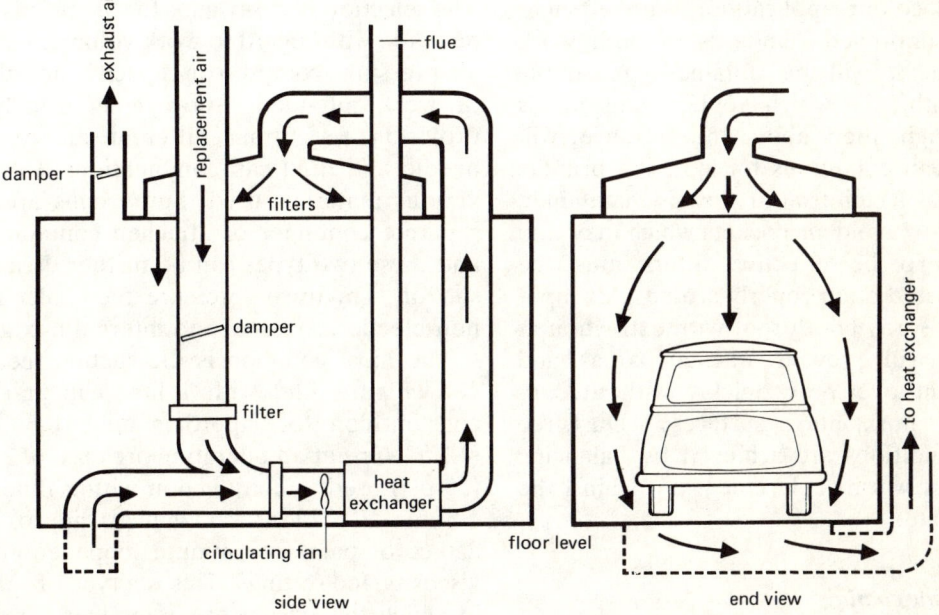

Figure 127 *Low-bake stoving oven*

Combined spray booth/oven

The spray booth unit is fitted with dual-speed fans and additional heating capability, so that after spraying the vehicle remains in the booth for the stoving or force drying cycles when the air flow from the ceiling to the floor is recirculated and increased to a minimum velocity of 30 m/min (100 ft/min) (see Figure 127). The unit then provides a rapid drying system for 1–2 vehicles per hour where uniform force drying or low bake improves the quality, durability and curing of finishes which would normally be air dried, irrespective of climatic conditions. A typical stoving schedule is 30 minutes at 60–80 °C. The advantages of the combined spray booth and oven are lower installation costs, a reduction in floor space requirements and reduced running costs when compared with a separate oven and a spray booth.

Open-ended booth

Open-ended spray booths or canopies may be used as areas for spraying primers prior to flatting and colour application. With efficient extraction, improved standards of both work and cleanliness will be obtained. A simple 'tunnel booth', in which replacement air is drawn through filters above the entrance, will establish clean conditions for spraying primers provided that a uniform air flow is maintained over the car to avoid air pockets which may trap overspray. To be effective filters must be securely located and properly sealed. Air input through the spray booth roof with extraction at floor level will prevent stagnant areas and remove paint overspray quickly without contaminating horizontal surfaces. Dust-free spraying conditions are achieved by balancing the supply of warm replacement air against the extraction rate.

Peelable booth coating

Cleanliness and freedom from dust and overspray is most easily achieved and maintained by utilizing a peelable booth coating on the exposed internal walls. The coating is applied by spray and removed when soiled by peeling directly from the surface to which it has been applied, thus ensuring that walls retain the appearance required for daylight conditions.

Respirators

Spray booth operators must be equipped with air-fed respirators, the purified breathing air supplied from a suitable purpose-built filter unit and regulator which complies with relevant British Standards. Airline respiratory equipment must only be connected to a regulated air supply.

All refinish operators must be supplied with a dust filtration mask for general working conditions, and an approved canister-type mask, particularly for sanding operations. Filter units must be maintained regularly.

Spray guns

The selection of spray guns for the refinisher will be made with regard to work volume, materials, air pressure, compressor capacity, and the type of work intended. Spray guns usually have provision for fitting different air cap, fluid needle and fluid cap combinations, to suit the various materials used. Spray guns are of the separate container or attached container types and these two types can be further divided into suction, gravity or pressure feed, bleeder and non-bleeder, external and internal mix guns.

The most common is the suction feed spray gun (Figure 128) with 1 litre cup and an air consumption of approximately 0.34 m^3/min (12 ft^3/min) at an operating pressure of 3.85 bar (55 psi) used in conjunction with a fluid cap of 1.8 mm (0.070 in) to deliver approximately 250 cc of paint per minute, depending on the viscosity and settings. This is a type of spray gun in which the stream of compressed air siphons material from the container attached to the

Refinishing facility and equipment 115

spray gun. Suction feed guns are used where there are many colour changes and they are the most popular in the car refinishing industry.

Gravity feed

The gravity feed gun (Figure 129) is suitable for spraying small quantities of viscous material. It is fed by force of gravity from the paint cup attached to the top of the gun body. The gravity feed gun uses the same air cap and fluid nozzle combination as the suction feed gun.

Pressure feed

This is a spray gun to which the paint is forced by pressure from a tank, fluid cup or pump (Figure 130). The gun is fitted with an air cap and fluid tip combination that does not siphon the paint and the fluid tip is generally flush with the air cap. A pressure feed gun is used when large

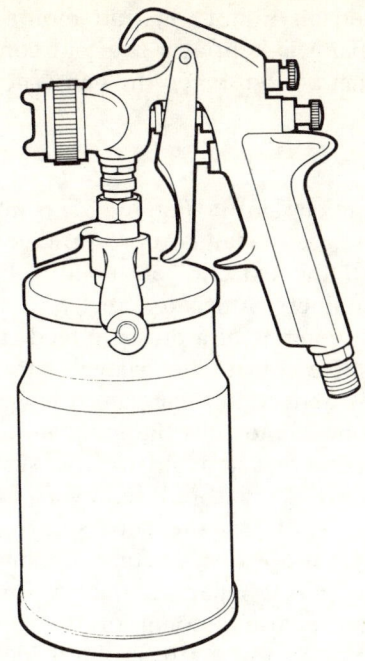

Figure 128   *Suction feed spray gun*

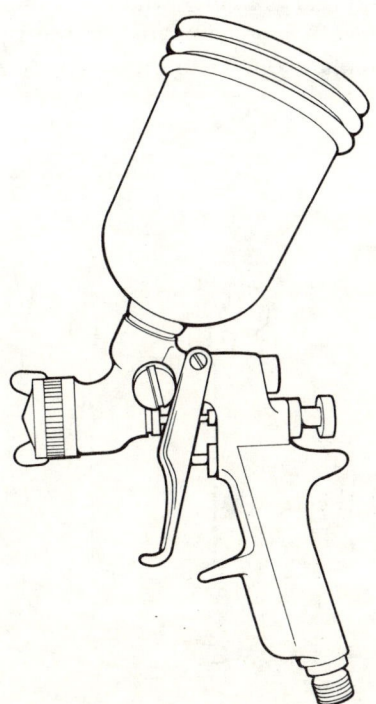

Figure 129   *Gravity feed spray gun*

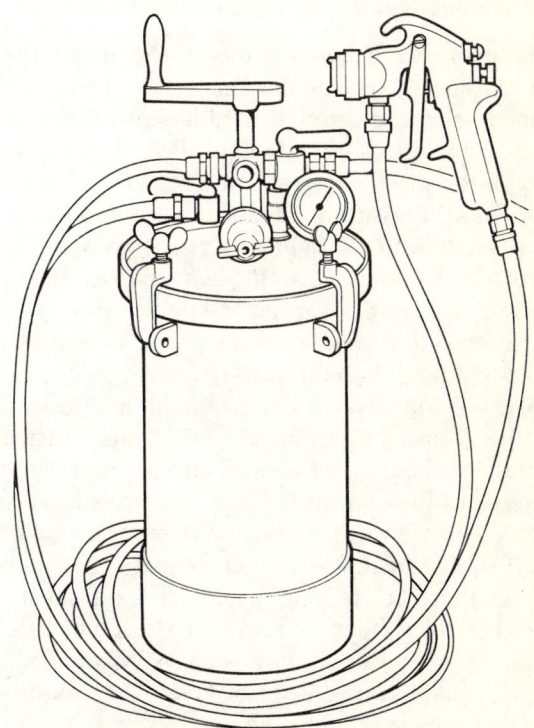

Figure 130   *Pressure feed spray equipment*

amounts of the same colour material are being sprayed, when the material is too heavy to be siphoned from a cup or container by suction, when extra fast application is required, or when large amounts of work need to be done very quickly, particularly on commercial vehicles. In all cases where large amounts of the same material have to be applied, the use of a pressure feed tank is recommended for the following reasons:

1. A very large amount of work can be done before filling is necessary, thus obviating the waste of time that would be entailed in constantly filling the smaller container.
2. The spray gun may be turned to any angle to coat the work effectively.
3. The material is fed to the gun in greater volume than any other method, particularly if heavy paints are used.
4. Less air pressure is required to obtain speed of operation.
5. Waste of paint and losses by evaporation are eliminated.

The principle of pressure feed is the application of low air pressure on the material in the tank so that it is forced through fluid hose to the spray gun. Air pressure is controlled by an air regulator on the lid and a pressure gauge is provided. Pressure feed tanks are in many cases provided with a light insert container that greatly facilitates cleaning and change of material; some tanks can be mounted on a castor base, making removal from place to place an easy matter. The tanks are strongly constructed to avoid any risk of distortion under pressure and are usually galvanized inside and outside. The lid is held on by clamps and is fitted with a gasket to prevent air leakage. There is a safety valve, an air release valve and at least one air and one fluid draw-off cock, but on the larger tanks there are two or three additional air and fluid cocks, so that the tank can be used by more than one operator. Provision is made for a hand-operated agitator to keep the material properly mixed and for large-capacity tanks it is advisable to have an agitator driven by a compressed air motor. The air motor ensures that the paint is kept at a constant consistency even during periods when the gun is not used.

Remote cup

The remote cup outfit (Figure 131) combines all the advantages of standard pressure feed equipment with the extreme portability of smaller paint containers attached directly to the gun. The outfit consists of a pressure feed cup fitted with sensitive controls to balance air and fluid flow. It is connected to the gun by lengths of air and fluid hose and thus the gun can always be held at the correct angle to any surface for uniform coverage and reduced overspray.

Pressure feed (remote cup) spray guns are high-speed set-ups with air consumption rates up to 0.74 m^3 (26 ft^3) per minute. A small fluid nozzle is used on account of the high paint discharge rate resulting from the 0.56–1.05 bar (8–15 psi) pressurization of the feed cup.

The fluid nozzle selected should be of the smallest internal diameter which has the capacity to pass sufficient paint to give the required speed of application. Then the air cap should be

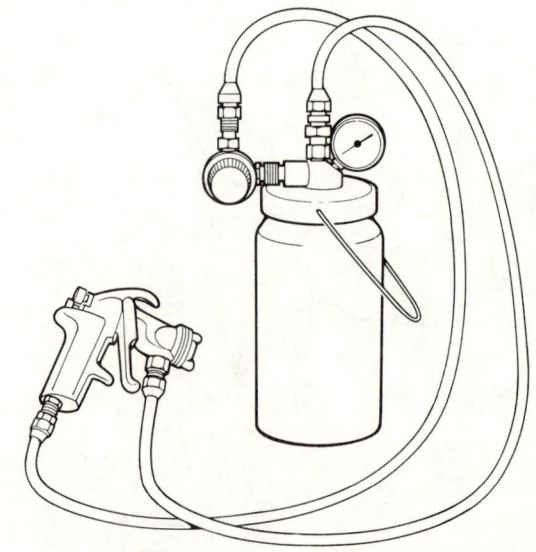

Figure 131 *Remote cup*

selected to give proper atomization. Details of the fluid nozzle/air cap combinations are given in the spray gun manufacturer's literature.

Spray gun construction

The basic functions of the main working parts (Figure 132) of a typical spray gun are:

1. The trigger, the action of which not only releases air into the air cap but also withdraws the needle from the fluid tip, thus allowing the paint to flow.
2. The fluid adjustment screw, which controls the amount of travel on the fluid needle to let more or less paint through the fluid tip.
3. The air valve, which is opened by the action of the trigger.
4. The spreader adjustment valve which controls the air to the holes in the horns of the air cap and regulates the size of the spray pattern from maximum width down to a narrow or round pattern.

The set-up

The phrase 'set-up' (see Figure 133) is a very important one in spray painting and includes three items which can be called the principal parts of the spray gun. They are the air cap, the fluid tip and the fluid needle. These are the parts

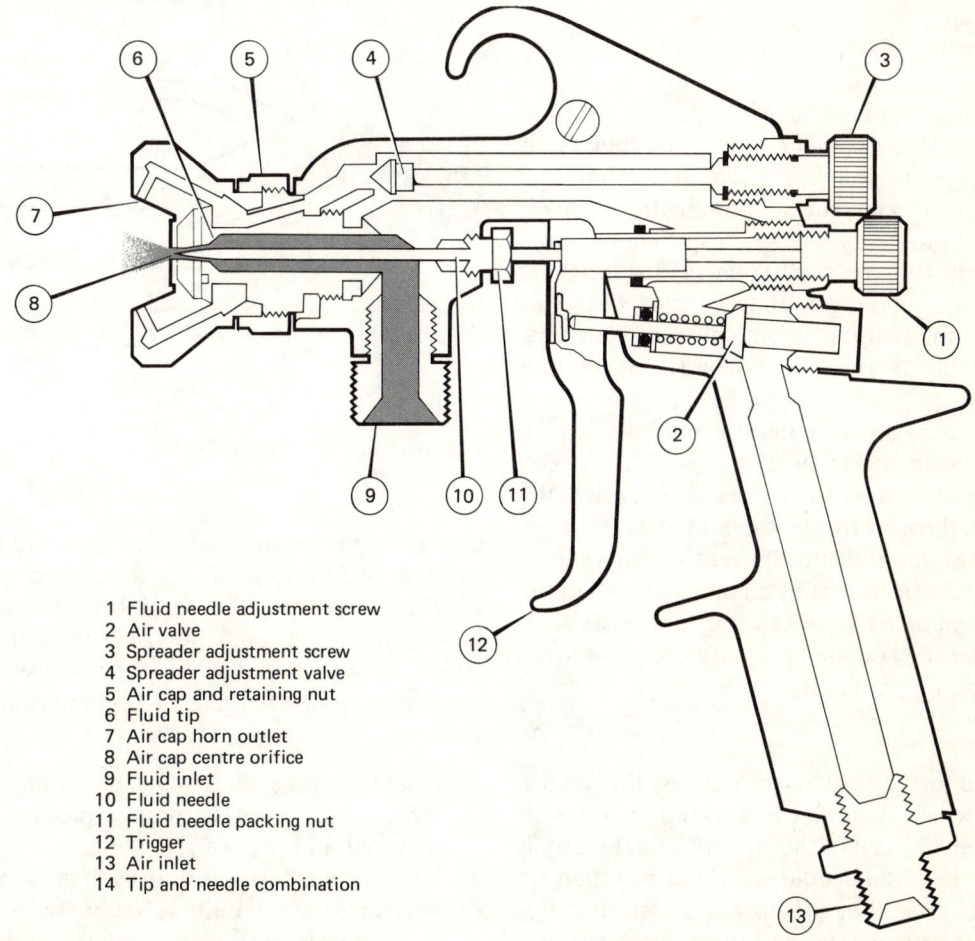

1 Fluid needle adjustment screw
2 Air valve
3 Spreader adjustment screw
4 Spreader adjustment valve
5 Air cap and retaining nut
6 Fluid tip
7 Air cap horn outlet
8 Air cap centre orifice
9 Fluid inlet
10 Fluid needle
11 Fluid needle packing nut
12 Trigger
13 Air inlet
14 Tip and needle combination

Figure 132 *Spray gun construction*

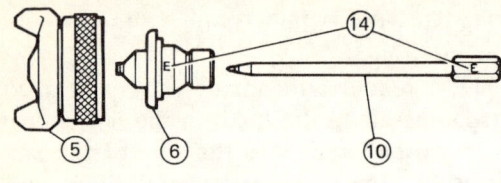

Note: See Key on Figure 132

Figure 133 *The set-up*

that are changed to make the spray gun more adaptable for applying one kind of material or another. They are changed according to whether suction or pressure feed is being used and to match the output volume of the air compressor.

The air cap
This atomizes the paint and forms the spray pattern.

There are two types of air cap. The external mix air cap (Figure 134(a)) has a number of air outlet holes in it and the air and paint mix beyond the air cap causing atomization. This is the more commonly used air cap. The internal mix air cap (Figure 134(b)) has a slot in it and the air and paint mix behind the slot thus atomizing the paint inside the air cap. This type is used with low-pressure, low-air-consumption guns.

The air caps are specified by a number which is stamped on the front of the cap. The spray gun, of course, uses air; it does this because the air passes through the air cap and after doing its job escapes into the atmosphere. The amount of air used (or consumed) by an air cap is called the air consumption of the cap and is measurable. The higher the spraying pressure the greater the volume of air required.

The fluid tip
The fluid tip (see Figure 133) is the nozzle through which the paint is directed into the air streams coming out of the air cap. The fluid tip is a seat for the fluid needle and these two then act as a valve controlling the flow of paint. Fluid tips are made in many sizes and the size reference is to the centre hole in the tip. When the gun is

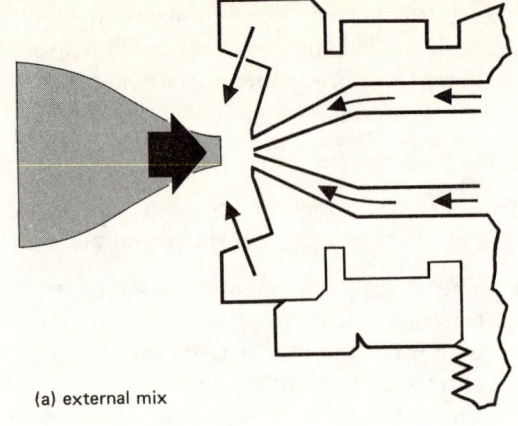

(a) external mix

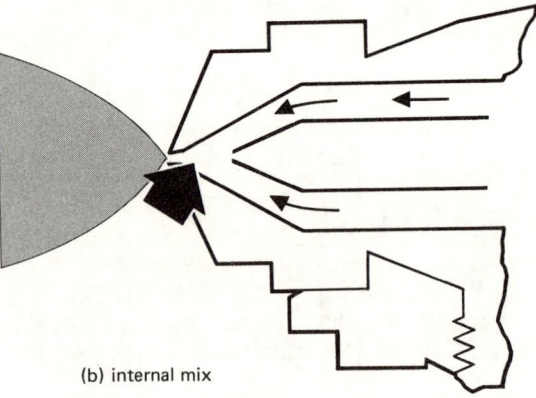

(b) internal mix

Figure 134 *The air cap*

operated, the paint flows from the fluid tip and the rate of flow depends on the diameter of the hole or nozzle of the fluid tip, the pressure behind the paint, and the viscosity of the paint itself. In selecting the fluid tip consideration should therefore be given to several important factors:

1 Heavy, coarse or fibrous materials require large nozzle sizes to permit passage of the material and prevent clogging.
2 Viscous materials requiring high atomizing pressures are handled better through the small nozzle sizes which assure more complete atomization.

3 Very thin materials that sag readily are applied at low atomizing pressures with small nozzle sizes to prevent excessive material application.
4 Abrasive or corrosive materials must be handled with tips made of wear-resistant or non-corrosive metals.
5 The type of material feed to be used is important since the nozzle sometimes recommended for pressure feed will not be satisfactory for suction feed although the converse does not apply.

The fluid needle

The fluid needle (see Figure 133) is a long pointed valve that seats into the fluid tip. Under the force of a spring it is normally pushed tight into the fluid tip opening, but when the trigger is pulled the needle is withdrawn from the tip thus allowing the paint to flow. Fluid needles are made in as many sizes as fluid tips to coincide with them and the needle must always be of the same size as the fluid tip.

It is important to remember that there is not just any combination of these parts that will do; they must be teamed together correctly for the type of paint to be sprayed, the surface to be covered, the amount of compressed air available and the speed permissible on the job. Very thin materials call for one combination, viscous tacky paints require another and very heavy coarse and fibrous materials still another. Thus the proper combination is essential if the best and most efficient performance is to be achieved.

Other equipment

General filters: air and paint filters

In-line miniature air filters may be fitted directly to the inlet tail on the spray gun and the air coupling attached to this so that any moisture is finally removed. Paint strainers are also available for fitment to the bottom of the suction pick-up pipe.

Gun pressure regulator

Spray guns may also be fitted with a thumb screw-type pressure regulator on the air inlet. Remote cup set-ups may also be fitted with a pressure indicator for both the cup pressure and the operating pressure at the gun.

Fluid coupling for remote cups

Quick detachable fluid couplings similar to an air coupling may be fitted to the fluid pipe between the gun and the cup to facilitate cleaning and refilling.

Hot spray cup

When spraying synthetic paints and commercial vehicle finishes, it is often advantageous to heat the paint as recommended by the paint manufacturer, prior to and also during spray application. A paint cup fitted with an electric heating element – a hot spray cup – will enable the paint to be heated and maintained at the correct temperature for application.

Small-capacity spray guns

Small-capacity spray guns are an asset for refinishing small areas and for local resprays, and particularly for door shuts etc. where overspray and the higher pressures required with a larger gun would otherwise be a problem or impair the finish when operating in close proximity to complicated body profiles.

Portable infra-red drying lamps

In smaller establishments where the volume of refinishing work or the type of repair is confined to single panels or smaller areas, portable infra-red drying units (Figure 135) are useful to increase throughput by cutting the drying times of primer coats and general minor repairs. Various sizes of unit are available to the refinisher to suit individual requirements.

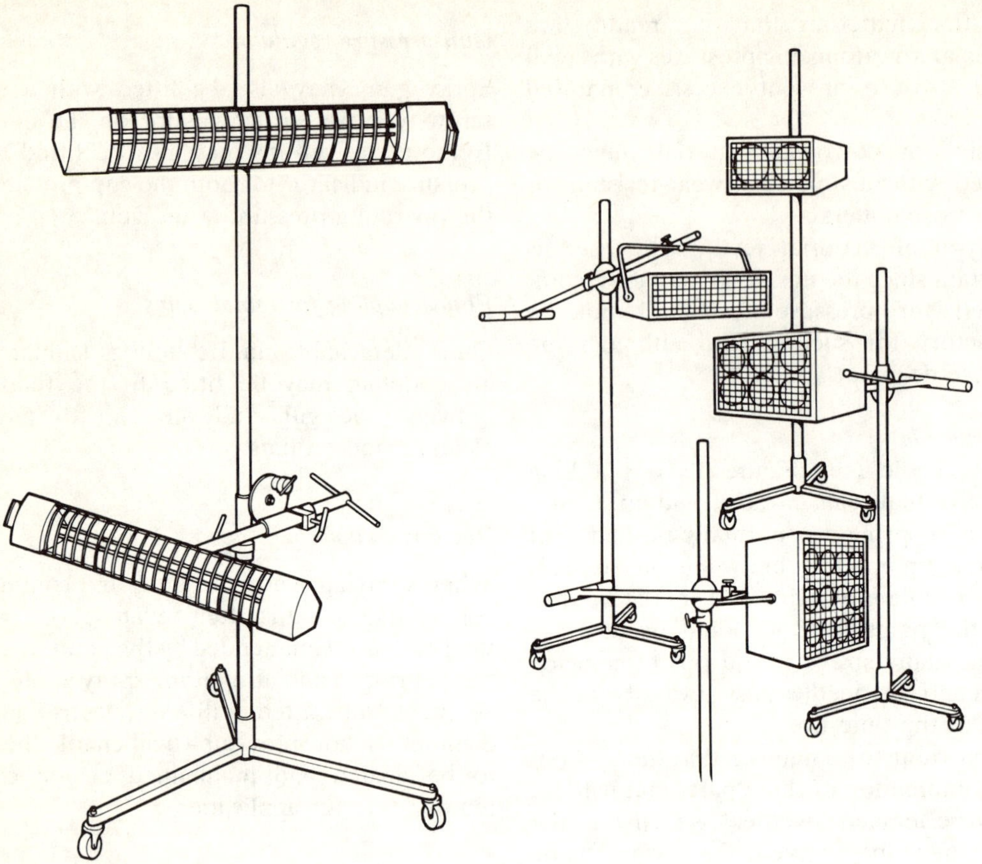

Figure 135 *Infra-red drying lamps*

Sanders

Note: operators must be equipped with a canister dust filtration mask when employed in any operation in which dust is generated. Refer to the relevant health and safety standards.

Various types of sander are available to promote fast, efficient dry sanding and wet flatting operations, and some of these machines incorporate an integral dust extraction facility.

Dual-action random orbital sander

Electric dual action (d.a.) sanders are heavier than the equivalent compressed air powered machines (Figure 136) and consequently are less popular. However, regardless of the type of

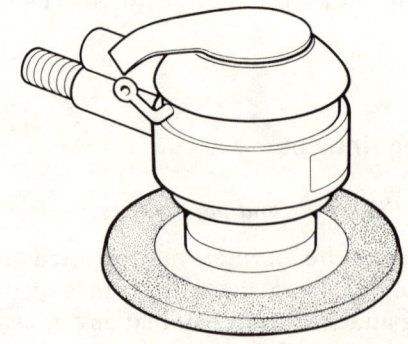

Figure 136 *Dual-action random orbital sander (air powered)*

machine, mechanical sanders are essential for feather edging and rapid preparation, and offer by far the fastest and most efficient means of dry sanding.

When operating, allow the machine to work correctly by applying only light pressure; excessive pressure on the machine will cause high machine and abrasive wear rates and rapid and premature dulling of the abrasive, and will result in a surface finish of poor quality. Do not run the sander without load or allow it to spin unnecessarily on the surface of the work. The average air consumption of a d.a. sander is approximately 0.255 m^3/min (9 ft^3/min) at 5.25–5.95 bar (75–85 psi). The general rule for machine sanding is that a finer surface finish is obtained relative to higher operating speed, and as these machines are often of high speed and exposed to dust, regular cleaning and proper lubrication is imperative.

Operate only at recommended air pressures consistent with proper performance. Select the abrasive grit size and paper disc, whether PSA (pressure-sensitive adhesive) type fitting or otherwise, to suit the machine, the surface and the desired preparation.

Type of backing pad The type of replacement backing pad fitted to the sander will have a pronounced effect on the life of the abrasive and the overall efficiency achieved. Self-adhesive paper discs used with a vinyl-surfaced backing pad will eliminate the possibility of inconsistent adhesion of the disc from non-uniform adhesive dispersion, which would result in rapid wear and poor economy of abrasive material and undesirable scoring of the paint surface.

Backing pads may be manufactured with a bevelled edge and from a flexible soft plastic, to facilitate use in awkward corners and on curved and complicated panel sections, etc.

Block sanders – reciprocating and orbital
Block sanders (Figure 137) are not usually of random orbit type but are constructed to provide a reciprocating or eccentrically orbiting rectangular pad, and are used in conjunction

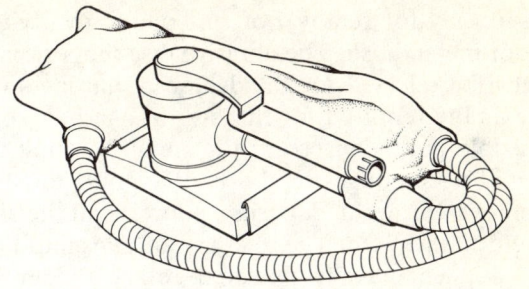

Figure 137 *Block sander with dust extraction (air powered)*

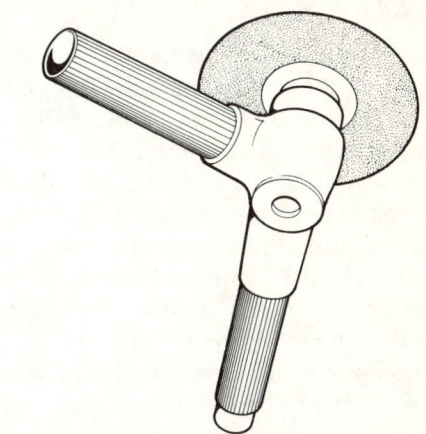

Figure 138 *Rotary action sander, disc type (air powered)*

with a waterproof abrasive for wet flatting or a stearate-impregnated abrasive paper for dry flatting. Speeds of operation vary, but this type of sanding machine is generally heavier and slower than the d.a. random orbital machines; it consequently produces a coarser texture, although it is commonly used for operations on surfaces that are flat or of large radius. The average air consumption is 0.200 m^3/min (8 ft^3/min) at approximately 5.95 bar (85 psi).

Rotary action sander – disc type
The rotary-type sanders are often electrically powered as these may be more efficient although heavier than their air powered counterparts (Figure 138). Rotary sanders are normally

used only for removal of rust or for rectifying poor metal finish. The abrasive disc support pad must be selected for the degree of hardness or flexibility required for the task in hand. A soft flexible pad is necessary for metal finishing to avoid deep gouges, and a hard support for the removal of weld deposits, poor metal finish, weld flashes, etc. The average air consumption of a pneumatically powered rotary sander is 0.255 m^3/min (9 ft^3/min) at 5.60–6.30 bar (80–90 psi).

High-speed polisher

Electric polishers are common items and, although heavier than compressed-air powered polishers (Figure 139), are more efficient in terms of energy consumption. The additional weight of an electric version may be a disadvantage during extensive polishing operations. The polisher may be of either orbital or rotary action and is generally used with lamb's-wool mops and foam heads for compounding, polishing and surface defect rectification work. The average air consumption of a pneumatically powered polisher is 0.255 m^3/min (9 ft^3/min) at 5.60–6.30 bar (80–90 psi).

Drill and rotary wire brush/abrading tool

A 9.5 mm (3/8 in) chuck drill is useful for removing stripper, cleaning paint, rust, etc. in conjunction with a suitable rotary wire brush or abrasive tool. Air powered drills are lighter and create less operator fatigue than equivalent electric drills but are less efficient in terms of overall energy consumption. The average air consumption is 0.170 m^3/min (6 ft^3/min) at approximately 5.95 bar (85 psi).

Tape and paper dispensers

Good-quality masking paper and tape should always be used as these will not tear easily or allow solvent penetration and can be efficiently dispensed from a tape and paper dispenser. High-quality tape allows masking before sanding

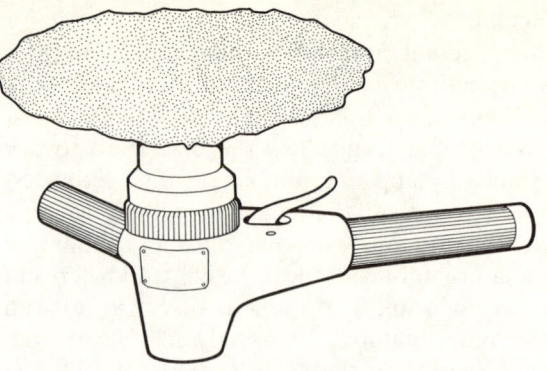

Figure 139 *High-speed polisher (air powered)*

operation and will resist curling from solvent action. Poor-quality tapes and paper often require reapplication after spraying primers, etc. Portable hand-held masking paper and tape dispensers are also useful items to facilitate the masking operation.

Disc dispensers

When self-adhesive sanding disc abrasives are used it is advantageous to use non-paper-backed discs from a roll and dispenser.

Viscosity cup

The flow cup, stand and stop-clock ensures paint application viscosity is correctly gauged. This equipment combined with its accurate simplicity is an essential part of the mixing and matching procedures.

Paint film thickness gauge

Paint film thickness measurement is essential if build-up problems are to be avoided. A simple magnetic 'pull-off' gauge will allow rapid assessment of film thickness. The rounded end of the magnetic portion of the gauge is placed in contact with and perpendicular to a clean area of the paint finish. As it is drawn away from the film the spring action in the gauge cylinder is opposed by the magnetic attraction of the vehicle body. The spring tension is dependent on the film thickness and this is indicated by a pointer and scale on the barrel of the instrument at the point when the magnet releases from the surface.

Compressed air blow guns

It is recommended that safety blow guns are always utilized; these are constructed with a secondary air jet adjacent to the working air jet, which issues an air envelop to deflect rebounding dust and dirt from the operator's eyes. The average air consumption is approximately 0.113 m^3/min (4 ft^3/min) at 5.95 bar (85 psi).

Basic refinishing rules

Following these basic rules is second nature to the good refinisher:

1. *Never skimp surface preparation* Removing wax, silicone, oil and grease with a water-miscible solvent cleaner, etching bare metal with an acid cleaner, and 'tack ragging' and 'petrol wiping' all surfaces immediately prior to painting, are all musts.
2. *Correct sanding technique* Remember that all sand scratches have to be filled by the subsequent painting operation. Eliminate sand scratch problems by using fine grit papers – 360/400 for surfacers, 500 for 'colour-on-colour' work – and by using the correct sanding technique – light pressure only, allowing the paper to do the work.
3. *Top-quality thinner* Economy on thinner is false economy. The saving per litre can be lost in many ways and will show up in greater usage and higher labour cost. Top-quality jobs go hand-in-hand with top-quality thinners.
4. *Correct amount of thinner* Always use the thinning ratio and viscosity specified for the product. Underthinning can cause slow drying and poor film appearance, overthinning can lead to poor durability. The use of a flow-type viscosity cup adds seconds to the job, but is an insurance against job rejects.
5. *Stir thoroughly* All paints need thorough stirring before use to ensure correct solids and correct colour.
6. *Correct atomizing pressure* Incorrect atomizing pressures can give rise to many paint faults. For example, too high a pressure can give excessive paint losses and poor 'flow', while too low a pressure can lead to 'popping', 'orange peel' and 'sags'. Do not overload the compressor and avoid pressure drop by using short airlines of large bore.
7. *Correct painting procedure* Keep the gun perpendicular to the work and 150–200 mm (6–8 in) from the surface. The spray fan should be 150–300 mm (6–12 in) wide and overlap on successive strokes about 50 per cent. Timing between coats is important to allow solvent to evaporate. With air-drying synthetic enamels, delay in second coat application, caused by a meal break say, can lead to wrinkling or lifting.
8. *Good ventilation* Plenty of fresh air above 15 °C should be available, especially during the drying stage. Acrylics, lacquers and synthetic enamels generally require 16 hours for complete drying. Never allow any finish to dry in a small closed booth or room.
9. *Good housekeeping* Keep the painting area clean, tidy, and safe. This is vital if the shop is to be smooth running and efficient and produce top-quality work.
10. *Maintenance of equipment* Keep all plant and equipment in good condition and properly adjusted. Renew any components at the first sign of wear. Top-quality work can only be produced from equipment in top-quality condition.

7 Spray painting principles

To make a success of spray painting, it is of vital importance that the operator first acquires the 'know-how', the technique of spraying, just as much as a brush-hand has to learn how to use his brush to get a first-class job done every time.

Naturally, the best painter by brush or spray is the man who has passed through all the basic training of his trade. It is not at all difficult for a beginner to learn spraying technique and a craftsman with painting experience will 'pick it up' very quickly indeed. It is merely a question of knowing how the spray gun works and how to use it, together with a good working knowledge of the air compressor and other equipment that goes with the gun.

Many leading spray painting equipment manufacturers will usually undertake to instruct operators in the use, care and maintenance of their equipment and the young painter is well advised to take every possible advantage of the manufacturer's advice, not only on the technique of spraying but also on the selection of the right equipment for his particular needs. The manufacturer's knowledge is the result of many years' research work in this particular field.

Mere words will never make a first-class spray craftsman out of a man who has never handled a spray gun before, for this is one of the arts about which it may be truly said that 'practice makes perfect'.

Flow cup and viscosity

Every painter knows that each kind of paint or finishing material has a different 'feel' or viscosity. Some flow out easily while some are very viscous or sluggish in their movement. The consistency or viscosity of a paint on application has a considerable bearing on:

1. The drying time.
2. The covering power.
3. The durability.
4. The general behaviour of the paint, depending whether the mix is too thick or too thin.

In addition the paint/solvent ratio or viscosity of a metallic paint can considerably affect the colour. To check paint viscosity, a flow cup or viscosity cup and stop-watch or clock must be used. The cup is simply an open-top container with a graduated outlet hole in the base (Figure 140).

There are two types of flow cup in common use. In each type there is a limited range denoted by a number which is stamped on the cup. Since the outlet hole sizes differ between types, and also according to the number, it is important that only the cup specified by the paint supplier is used to determine viscosity.

Use of flow cup

Use of the flow cup is quite straightforward, but it is important that the paint is at shop temperature (15–21 °C). The flow cup is held or placed in a stand in a level position; a finger is placed over the outlet hole and the cup is then filled until paint just flows over into the spill rim at the top. Remove the finger from the outlet hole and simultaneously start the stop-watch, allowing the paint to flow into a clean container. The time taken for the cup to empty is indicated by the first break in the stream of paint issuing from the outlet hole. Immediately the paint stream breaks, stop the watch and the time indicated is the viscosity of the material. The cup should be cleaned at once after use and carefully stored so that the outlet hole is not damaged or enlarged in handling. A good spray gun can quite easily produce a perfectly good finish with any of the wide variety of paints available, provided that the correct spray gun adjustments are made to suit the material and provided, of course, that the paint itself is of reasonably good quality.

Spray gun adjustments

The all-important spray gun adjustments are as follows:

1 The volume flow of paint from the nozzle to suit the size of the area to be covered and the speed of operation required.

2 The correct proportioning of atomizing air pressure to the flow of material.
3 The proper spray width adjustment.

All these adjustments are interdependent as they vary according to the viscosity of the paint, the volume and pressure of compressed air available, and the sizes of areas to be covered.

The correct spraying pressure for any paint depends primarily on the paint viscosity and spray gun set-up. With a modern, fast, suction feed set-up, cellulose lacquers and synthetic enamels should be sprayed at a higher air pressure than acrylic lacquers in view of the lower application viscosity of the latter. Spraying pressure should never exceed that required to obtain proper atomization, as this will only result in excessive paint loss and poor flow owing to high solvent evaporation before the paint reaches the surface. On the other hand, too low a spraying pressure will give films with poor drying characteristics and, because of high solvent retention, prone to bubbling and sagging.

Spray patterns

The normal spray pattern (Figure 141) is a rectangle with rounded ends. The paint must be distributed evenly throughout the pattern and there must be no distortion of the shape.

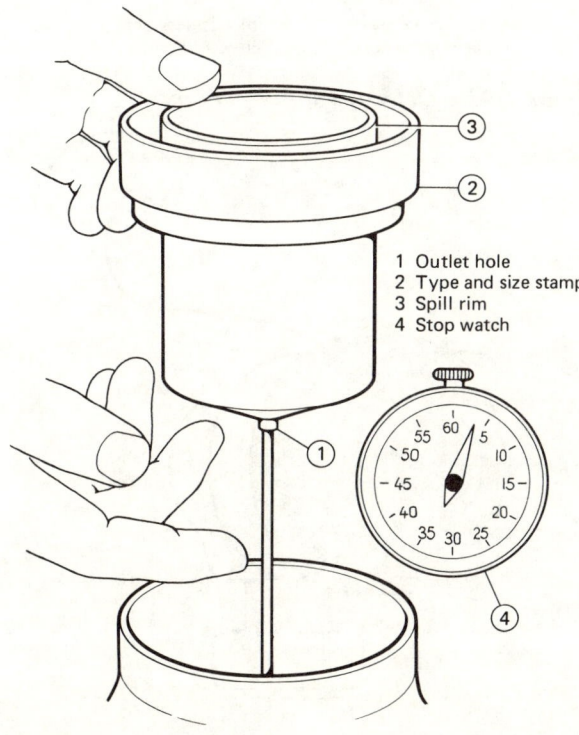

1 Outlet hole
2 Type and size stamp
3 Spill rim
4 Stop watch

Figure 140 *Flow cup and viscosity*

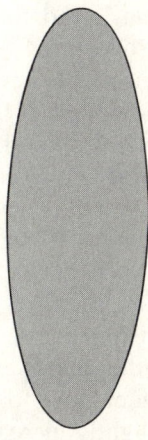

Figure 141 *Normal spray pattern*

Reduction of the pattern width to a round spot must be accompanied by a reduction in paint delivery or overloading will result. Pattern size is governed by the type and capacity of spray gun set-up and may be adjusted by controls on the spray gun. It should not be attempted to vary the pattern size by increasing or decreasing the distance between the spray gun and work surface.

Spray pattern test

With the paint reduced to the correct spraying viscosity, the atomizing air pressure adjusted and the gun connected to the airline, the gun should be set by opening the spreader adjustment as far as it will go and adjusting the fluid screw until the first thread of the screw is visible.

A test pattern can then be sprayed onto waste material to check for uniform paint distribution. This is done by holding the gun steady and momentarily triggering on and off with the gun approximately 150–200 mm (6–8 in) from the surface. Then a small area is sprayed to check the speed of operation. If the spray pattern appears starved of material, the fluid adjusting screw is opened wider to allow more paint through. If too much paint is applied the paint flow is reduced by screwing in the fluid adjusting screw or by reducing the air pressure at the transformer.

If the atomization is too fine (this is recognized by excessive overspray or dry spray) the atomizing pressure is reduced whilst keeping the fluid adjustment wide open. If the atomization is too coarse (recognized by a speckled effect or dimple finish lacking flow-out of paint) the atomizing air pressure is increased or the material flow cut down. If, when the spreader valve and fluid screw are wide open, the spray pattern is too narrow, the fluid flow is increased by raising the atomizing pressure or by thinning the material. A wetter coat may be obtained by turning in the spreader valve to narrow the spray pattern and then slightly increasing the atomizing pressure.

Defective spray patterns

If dirt or dry paint become plugged into the horn holes of the air cap, on the top of the fluid tip, or on the air cap or fluid tip seat, a defective spray pattern will be obtained and this should be

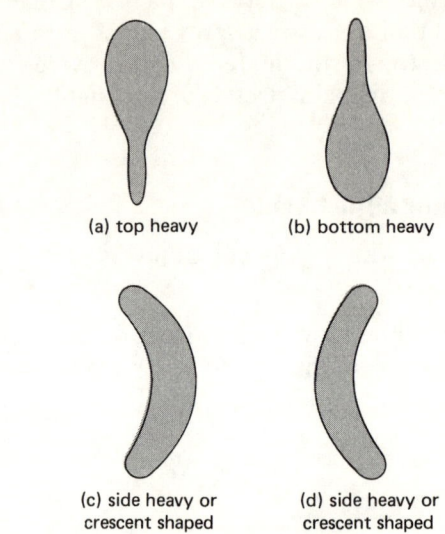

(a) top heavy   (b) bottom heavy

(c) side heavy or crescent shaped   (d) side heavy or crescent shaped

Figure 142 *Defective spray patterns*

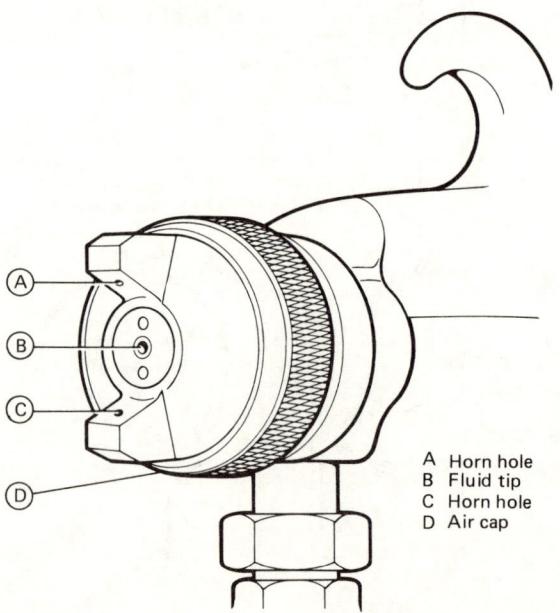

A Horn hole
B Fluid tip
C Horn hole
D Air cap

Figure 143 *Correction of defective spray patterns*

rectified by careful cleaning. A defective spray pattern will show when a test trigger pull is made and the shape of the pattern indicates where to look for the obstruction.

For any of the defective patterns in Figure 142(a)–(d), first rotate the air cap (D in Figure 143) one half-turn and spray another pattern. If the defect is then inverted, the obstruction is in the air cap. Clean out the horn holes (A and C in Figure 143). If the pattern is not inverted the obstruction is on the fluid tip (hole B). Whilst the air cap is easily cleaned, the defective spray pattern may be caused by a fine burr on the edge of the fluid tip or by dried paint just inside the opening. In the former case it can be removed by fine wet or dry sandpaper and in the latter by cleaning.

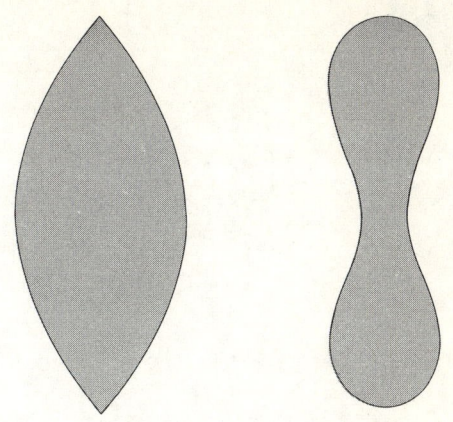

Figure 144 *Heavy-centre spray pattern*

Figure 145 *Split pattern*

The heavy-centre incorrect spray pattern
This pattern (Figure 144) is caused by:

1 Setting the spreader adjustment valve too low.
2 Atomizing air pressure too low or paint too thick.
3 Nozzle too large for the paint.
4 Nozzle too small.

Sometimes when pressure feed is used a spray pattern is split owing to the air and fluid pressures not being properly adjusted (Figure 145). This can be put right by reducing the width of the spray pattern by means of the spreader adjustment valve or increasing the fluid pressure, adjusting atomization as necessary. It should be remembered that this latter adjustment increases speed and the gun must be handled much faster.

Spraying technique

Spray painting is a craft, as is any other method of painting, but by the knowledge and correct use of spray gun technique a surface can be given a first-class coating in the minimum amount of time, with greatly reduced paint wastage and less fatigue for the operator. Many painting jobs have been spoiled by lack of knowledge of this technique or by disregard of its value to the painter, and it cannot be too strongly emphasized that a little time spent in study and practice will be amply repaid by the greatly improved quality of finishes obtained as well as by the saving of time and material.

The following are basic rules of spraying paint; and it is assumed that the operator has the right gun for the job, has mixed and strained his paint correctly, and has adjusted the fluid and atomizing air pressures to suit the material and speed of operation required.

The gun should be held at right angles to the surface to be painted. The distance between the surface and the face of the air cap should be 150–200 mm (6–8 in), or a hand span as a quick check (Figure 146). At this distance it will create its best atomization and distribute the fluid coming out of the gun to the full width of the pattern size.

Gun stroke triggering

Each stroke is made with a free arm motion across the face of the surface with the wrist flexible so that the gun is kept at right angles to the surface and at the correct distance from it

128 Principles and Practice of Vehicle Body Repair

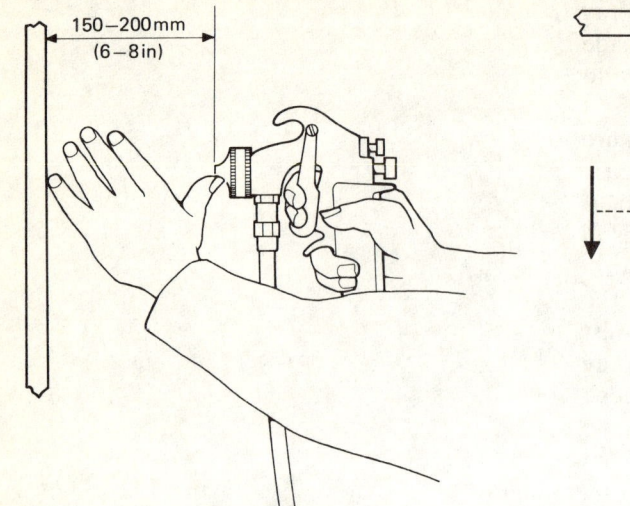

Figure 146 *Spray gun distance*

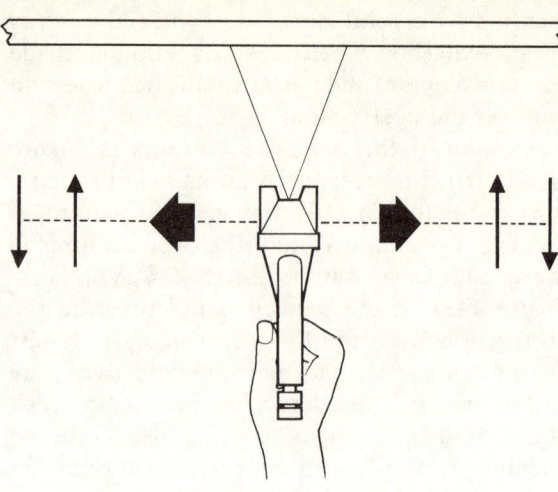

Figure 147 *Correct gun stroke triggering*

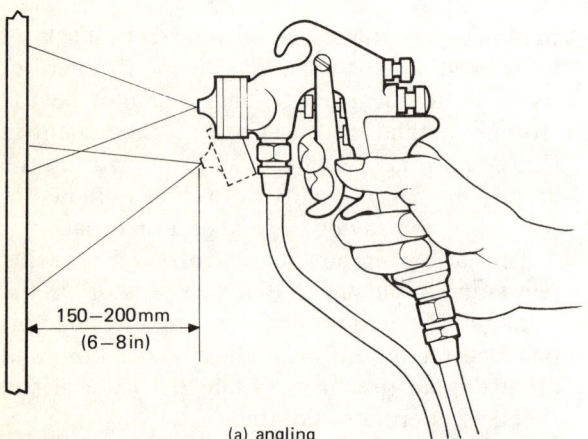

(a) angling

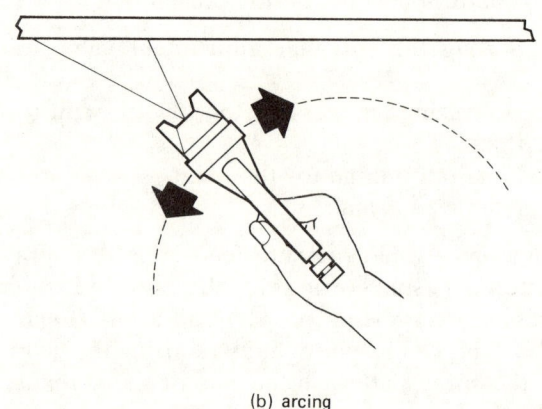

(b) arcing

Figure 148 *Incorrect gun–surface relationship*

(Figure 147). The speed of each successive stroke must be constant so as to maintain a uniform thickness of coating. Correct triggering of the gun is also an essential part of stroke technique. This means that, in order to prevent the building up of paint on the surface at the beginning and end of each stroke, the movement of the gun should be started before the trigger is pulled and the trigger released again before the gun movement ceases at the end of the stroke. It is only necessary that the flow of paint be stopped. Upon developing a feel or touch for the trigger it is accepted practice to permit the flow of air through the gun at all times, stopping only the paint. This also allows a slight increase in speed. Failure to trigger the gun will result in an increase of overspray and may cause sags, runs and streaks.

Gun-surface relationship

To ensure a uniform deposit of paint on the surface, it is imperative that the gun be stroked across the surface at a right angle to it. Pointing the gun to either side (arcing) or up and down (angling) will result in a non-uniform deposit of paint, giving an uneven spray pattern (Figure 148). The wrist should be kept flexible during the spraying process. In certain types of work it is often necessary to tilt the gun but this should not be done on surfaces suited to the correct gun position. Normally the stroke is made in a horizontal direction with the object being sprayed. If a vertical stroke is preferred or is necessary, the operator need only loosen the air cap retaining ring and rotate the air cap 90 degrees; this gives a horizontal fan pattern and will facilitate vertical painting.

Overlap

The edges of a spray pattern taper off slightly and to obtain even coverage of a surface it is essential to overlap the previous stroke by approximately 50 per cent (Figure 149). This eliminates the need of a double or cross coat and will assure the deposit of a wet coat without the risk of streaks or patches. It is a good plan to aim the gun at the extreme edge of the previous stroke to ensure getting the required overlap. Long work is sprayed in sections of convenient length, each section overlapping the previous section by 100 mm or so. The spray gun trigger should always be handled smoothly and not pulled abruptly or released with a jerk.

Spray gun maintenance

Cleaning spray painting equipment is an easy operation, and provided it is done systematically and thoroughly after every job is finished it will pay dividends in better spraying and trouble-free gun performance. One should develop the habit of cleaning equipment promptly after spraying to avoid the possibility of having to remove hardened paint, particularly from the fluid tip and passages of the spray gun. The cleaning fluid used must, of course, be suitable for the kind of paint that has been sprayed – water for water colours, methylated spirits for spirit colours, turpentine or white spirit for oil colours and thinners for cellulose or synthetics. A caustic solution must never be used for cleaning a spray

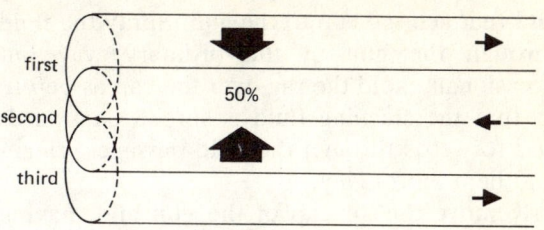

Figure 149 *Spraying overlap*

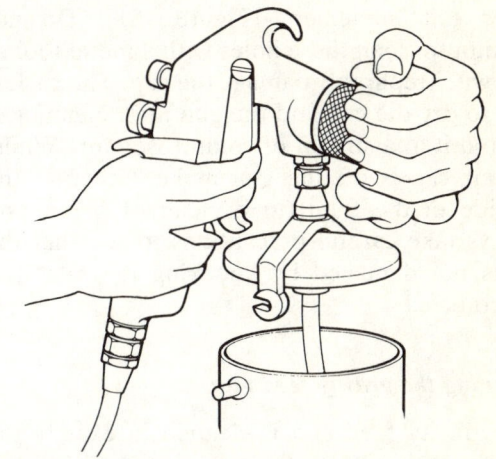

Figure 150 *Suction feed gun cleaning*

gun or any other parts of the equipment as it will inevitably attack the metal of which they are constructed. Also, the spray gun must not be immersed in cleaning fluid which will destroy the lubricant in the fluid needle and air valve packings.

Cleaning the suction feed gun

After spraying, and whilst the gun is still connected to the compressed air supply, loosen the cup. With the fluid tube still within the cup, hold a piece of rag lightly over the centre hole of the air cap. Pull the trigger and the rag pad will turn back the compressed air through the fluid tip, thus forcing any surplus paint in the gun and fluid tube back into the cup (Figure 150). Empty the cup, allowing it to drain for a few moments partially refill it with a suitable cleaning fluid, and reattach the cup to the gun. Spray the fluid through the gun in the ordinary way but occasionally hold the rag over the cap as before so that the cleaning fluid is surged backwards and forwards through the fluid passages, cleaning them thoroughly.

Remove the air cap of the gun and, having soaked it in cleaning fluid, scrub it with a stiff brush. If any of the holes in the cap are clogged, probe them with a matchstick, toothpick or other soft implement (Figure 151). Do not attempt to clean these holes with a metal tool as that will irreparably damage the cap. The easiest way to dry the cap and the gun after cleaning is to hold it in a stream of compressed air. Whilst the air cap is off the gun make sure that the outside of the fluid tip is clean of paint, and always take careful precautions to see that the tip is not damaged by knocking it while it is unprotected.

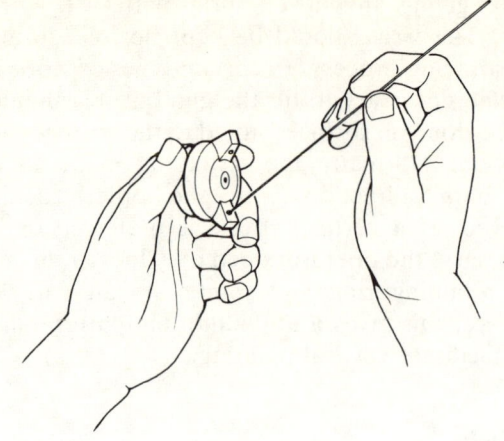

Figure 151   *Air cap cleaning*

Cleaning the gravity feed gun

The gravity feed gun is cleaned in exactly the same way except that the cup is not detached but the lid is taken off. When blowing back with a gravity feed gun special care should be taken to ensure that the cup is turned away from the face as a certain amount of cleaning fluid is liable to be blown out of the cup.

In the lids of both gravity and suction feed cups there is a vent hole to allow air to enter the cup to replace paint that is drawn out, and it is essential that this vent hole is kept open as otherwise the paint will not flow out of the gun properly. There is also a lid gasket which must be carefully cleaned to ensure that it functions properly.

Cleaning the pressure feed system

To clean the pressure feed system, turn off the cock supplying compressed air to the pressure feed tank and open the relief valve on the tank lid. This will release the air pressure within the tank. After this has been done, and not before, unscrew the clamps that hold the tank lid down.

Lift the tank lid and replace it slightly out of position so that atmosphere air is free to enter or leave the tank. Now turn the air cock on again and force the paint in the fluid hose back into the tank by holding a rag over the spray gun air cap (Figure 152).

Empty and drain the paint from the fluid tank and replace it with cleaning fluid, which is sprayed and blown back until the hose and fluid passages are clean. Detach the hose from the gun and tank and hang it up on a suitable peg with both ends over a receptacle into which it can drain.

Clean and lubricate the gun; also clean the tank inside and out including the lid gasket. The air pressure regulating valve on the tank should be unscrewed until it feels free so as to relieve the tension on the valve spring.

When the pressure feed equipment is to be used again in a few hours it is unnecessary to clean it, but the parts should be left connected with the fluid hose full of material and the lid of the pressure feed tank tightly in place.

Cleaning other equipment

Other equipment besides the spray gun and tank should always be kept scrupulously clean to ensure their constant efficiency. The air hose

Spray painting principles 131

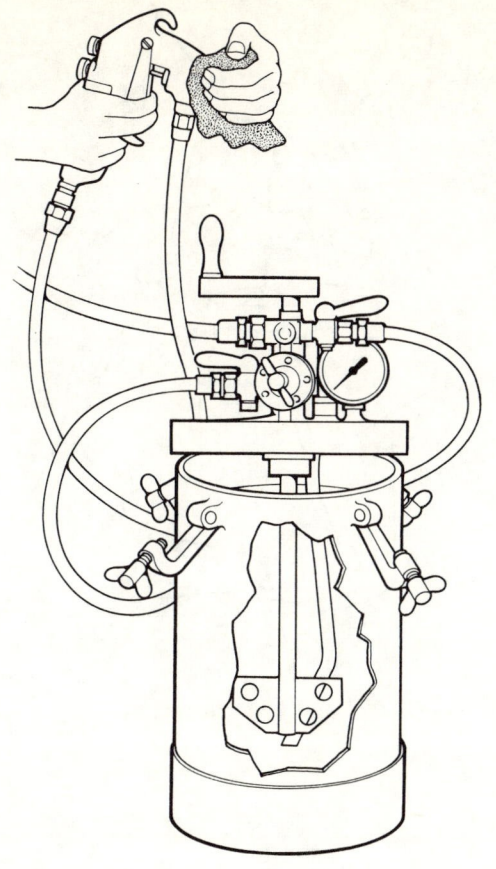

Figure 152 *Pressure feed system cleaning*

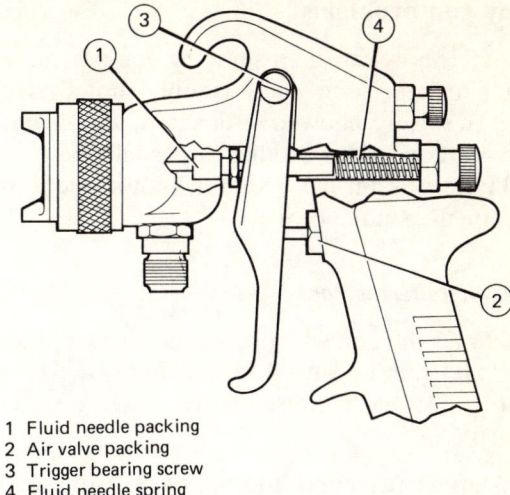

1 Fluid needle packing
2 Air valve packing
3 Trigger bearing screw
4 Fluid needle spring

Figure 153 *Spray gun lubrication*

should always be blown out with compressed air before it is attached to the gun, so that any dust or other loose material that may have lodged in it is removed. Keep the outside of the fluid hose free from accumulation of material; the life of the hose depends upon the care given to it.

Another important point is to keep hose connections clean and undamaged, if they cannot seat properly leakages will occur at the connection, causing waste of air and paint.

The oil level in the air compressor sump should be inspected regularly and, if necessary, topped up with the oil recommended by the manufacturers. The sump should also be emptied periodically and fresh oil put in, the length of this period depending on the amount of work the compressor has been called upon to do.

Another feature of the air compressor is the air inlet which is usually provided with a filter. This filter must be kept scrupulously clean and the filtering medium within it replaced when it becomes laden with dirt.

The air receiver is fitted with a drain cock which should be opened at least once a day, with the compressor running, to blow out accumulated water and oil. When the air compressor is fitted with an air transformer the drain cock of the transformer must also be opened at least once a day. If this is not done, the transformer will become waterlogged and moisture will get into the air hose.

Spray gun lubrication

At least once a week the spray gun should be lubricated by placing a drop of oil on the fluid needle packing, the air valve packing and the trigger bearing screw (Figure 153). The fluid needle packing should be removed occasionally and softened with oil and the fluid needle spring should be coated with vaseline.

132 Principles and Practice of Vehicle Body Repair

Spray gun problems

No matter how excellent spraying equipment is, sooner or later some small trouble shows itself which, if it were allowed to develop, would mar the work done. This trouble can usually be very quickly rectified if the operator knows where to look for the source of it.

Jerky or fluttering spray

Sometimes the gun will give a fluttering or jerky spray and this is caused by the following (see Figure 154: each cause is indicated on the figure):

1 Insufficient paint in the cup or pressure feed tank so that the end of the fluid tube is uncovered.
2 When a suction feed gun is used, the cup is tilted at an excessive angle so that the fluid tube does not dip below the surface of the paint.
3 Some obstruction in the fluid passageway which must be removed.
4 Fluid tube loose or cracked or resting on the bottom of the paint container. To remedy this, the fluid tube should be carefully bent slightly upwards to clear the bottom of the cup.
5 A loose fluid tip on the spray gun.
6 Too heavy a material for suction feed.
7 A clogged air vent in the cup lid.
8 Loose nut coupling the suction feed cup or fluid hose to the spray gun.
9 Loose fluid needle packing nut or dry packing.

Paint leakage from front of the spray gun

This is caused by the fluid needle not seating properly, owing to the following (see Figure 155: each cause is indicated on the figure):

1 Worn or damaged fluid tip or needle.
2 Lumps of dried paint or dirt lodged in the fluid tip.
3 Fluid needle packing nut screwed up too tightly.

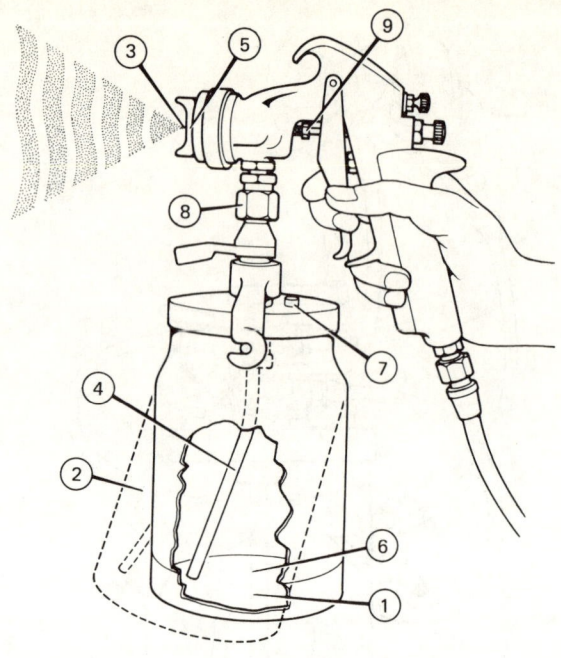

Figure 154 *Jerky or fluttering spray*

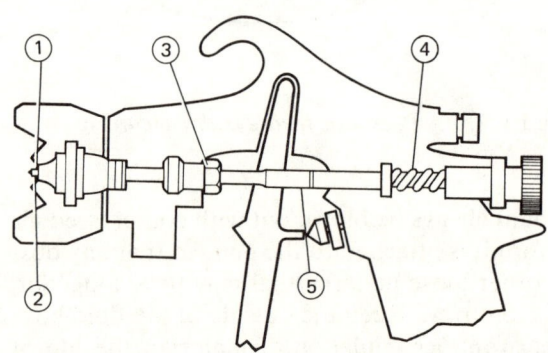

Figure 155 *Paint leakage from front of gun*

4 Broken fluid needle spring.
5 Incorrect size needle.

Paint leakage from the fluid needle packing nut is caused by a loose packing nut or a worn or dry fluid needle packing. The packing can be lubricated with a drop or two of light oil, but fitting a new packing is strongly advised. Tighten the packing nut with the fingers only to prevent leakage, but not so tight as to bind the needle.

Compressed air leakage from front of gun

Compressed air leakage from the front of the spray gun is caused by the following (see Figure 156: each cause is indicated on the figure):

1. Dirt on the air valve or air valve seating.
2. Worn or damaged air valve or seating.
3. Broken air valve spring.
4. Sticking valve stem owing to lack of lubrication.
5. Bent valve stem.
6. Lack of lubrication on air valve packing.
7. Air valve gasket damaged.

If the air compressor pumps oil into the air line, it is for the following reasons:

1. Strainer on air intake clogged with dirt.
2. Clogged intake valve.
3. Too much oil in the crankcase.
4. Worn piston rings.

An overheated air compressor is caused by:

1. No oil in the crankcase.
2. Oil too heavy.
3. Valves sticking or dirty and covered with carbon.
4. Insufficient air circulating round an air-cooled compressor owing to it being placed too close to a wall or in a confined space.
5. Cylinder block and head being coated with a thick deposit of paint or dirt.
6. Air inlet strainer clogged.

Motion study – spraying procedure

Once having mastered the spray technique it is necessary to learn the basic principles of motion study in order to obtain maximum efficiency in spraying with minimum of fatigue.

It is not sufficient just to spray correctly; the operator should be able to work out the quickest

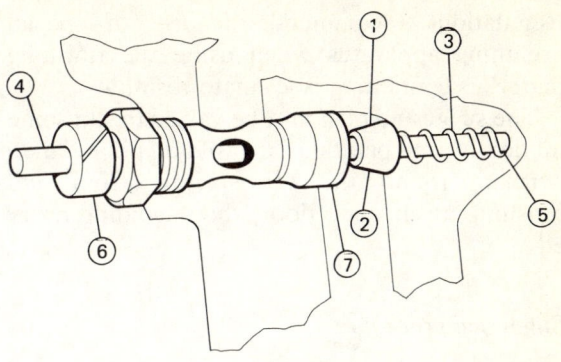

Figure 156   *Air leakage from front of gun*

way round a job so that the fewest number of gun strokes are necessary to cover the surface adequately, at the same time avoiding paint wastage and reducing the moving of equipment, trestles, ladders and similar items to a minimum. Never overreach in spraying but always confine the stroke to a comfortable length, usually about one metre or a little more. Overreaching inevitably means that the gun is angled, thus causing spray fog through the paint being deflected from the surface. It is advisable to practice spraying with either hand, as this will allow a larger area to be covered from one position and help to reduce fatigue by giving each hand a rest in turn.

Spraying plan

If a complete car respray is necessary, a painting sequence should always be planned before commencing to spray, primarily for the sake of continuity so that dry joint overlaps are avoided and unnecessary work is minimized. First paint all inaccessible areas such as boot and bonnet edges, channels and door shuts, with doors etc. held slightly ajar to prevent sticking and permit proper drying. This initial painting operation has the effect of laying the dust which might otherwise have blown out from these openings and spoilt the exterior finish work. In the United Kingdom current Health and Safety at Work

Regulations recommend the use of special breathing apparatus when using the finishing materials containing isocyanate resin.

The suggested plan may be varied to suit some models by first painting the roof and pillars, then working round the car starting from and finishing at an open door, thus avoiding a dry edge.

Suggested procedure

A suggested order of painting is as follows (see Figure 157: each stage is indicated on the figure):

1. Once the inaccessible areas have been painted, commence by painting the roof panel on one side, starting at the roof to rear quarter panel and drip rail to include the windscreen pillars. Paint from the drip rail in towards the approximate centre line of the panel.
2. Move to the opposite side of the car, pick up the wet paint edge by applying the paint from the centre of the roof out towards the drip rail to include the windscreen pillar, and finish off the roof by painting the roof to rear quarter panel.
3. Extend this application to coat the rear quarter panel on that side, then paint the entire door to include:
4. The doors and the front wing on that side.
5. Paint the front cowl, bonnet top and top of the opposite front wing.
6. Move round the car, paint the front wing.
7. Door(s) and
8. Rear quarter panel on that side.
9. Conclude by painting the rear cowl, boot lid and back panel.

Single and double coats

A single coat is simply one pass of the spray gun. A double coat consists of two passes of the spray gun and is usually accomplished by painting an area such as a wing or door, and then immediately painting it again before moving on

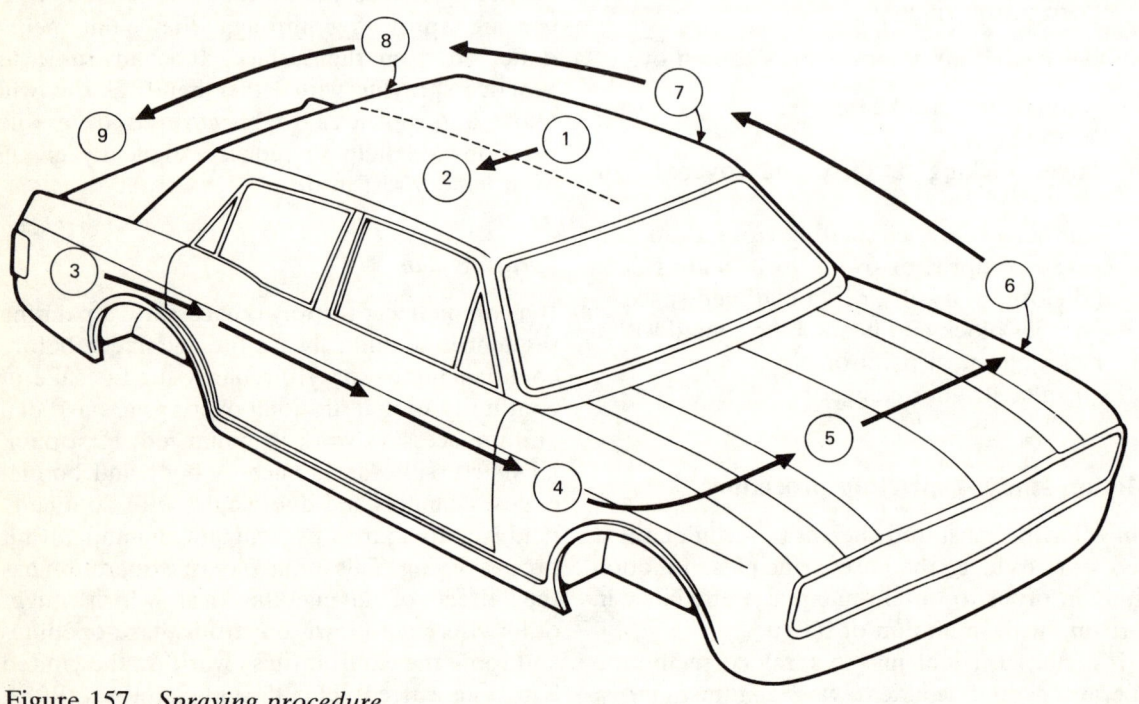

Figure 157 *Spraying procedure*

to the adjoining panel. Single coats give rapid solvent release and fast through-drying. Double coats give good flow, gloss, build and freedom from dryspray. For example, cellulose primer–surfacer may be built up using one or two single coats to give fast drying and minimize the risk of trapped solvent attacking or softening the underlying existing finish, followed by a double coat to give build and good flow, thus reducing the amount of flatting. Lacquer type finishes may be treated in the same way, applying thin single coats first to achieve maximum holdout and minimal solvent penetration down into undercoats and filled areas. The fast drying of the thin single coats prevents excessive dirt pick-up and these built-up colour coats will be hard and ready for finishing with a double coat in the shortest possible time. With synthetic enamels, a single coat is allowed to flash off until most of its solvent is lost and it becomes a 'holding' coat for a final full wet coat obtained by double-coat application.

Flash-off periods

Time must be allowed between the application of each coat of paint to allow the solvents to evaporate. Failure to allow adequate flash-off time will result in an excessive amount of solvent remaining in the paint system when the work is complete, and this can cause many paint failures. The paint manufacturer's instructions will apply to average conditions of working, but an allowance must be made when paintshop and/or climatic conditions are abnormal. For example, if conditions are cold and damp or if air movement is poor, the flash-off time must be increased. In hot dry conditions, or if air movement is high, a shorter period may be adequate. A thick coat applied at high viscosity will require a longer flash-off time than a thin coat of the same paint applied at low viscosity.

It is most important to allow adequate flash-off time when force drying with infra-red lamps or in a convection oven, as excess solvent in the paint film may boil if heat is applied too soon.

Paint heating

Paint heating is widely used for industrial spray painting because it provides the means to apply a much greater coverage per litre of paint than is possible by conventional cold spraying; it can also be profitably employed by the motor vehicle refinisher. When paints of many types are heated their viscosity is considerably reduced and, by adding a paint heater unit to standard spray equipment, the operator has complete control over the paint viscosity whatever the atmospheric temperature. He also has the major advantages of saving paint and time.

To get a smooth finish with any material when using the cold spray method it is essential to obtain fine atomization and for this purpose solvent is usually added to reduce the paint viscosity. The more solvent that is added, the thinner the paint will become with increased tendency to run or sag, apart from the more obvious disadvantages such as loss of gloss, depth of coat, colour, etc. It is clear, therefore, that only sufficient solvent should be added to produce fine atomization and film build at the spraying pressure. In the process of spraying, because of the impact speed due to the high air pressure used, there is a certain amount of rebound from the surface. This creates spray fog and a recognized loss of material. When cold spraying is done out of doors or in places where the ambient temperature constantly changes, the viscosity of the material also varies, requiring frequent adjustments to air and fluid pressures and viscosity. Uneven film build and persistent runs and sags are likely to result.

Low-pressure spraying with heated paint

By using the paint heating system with the same spray gun and material used for cold spraying, a 20 to 30 per cent or even greater saving can be made on the amount of paint required for a job. The method of making this saving is to thin the material exactly as for cold spraying and to heat it to approximately 70 °C, which reduces the paint viscosity still further. (To effect maximum paint saving a fast evaporating solvent must be used for thinning the material. If this is not the thinner normally used, the paint supplier should be asked to recommend a fast evaporating solvent suitable for the material to be sprayed.)

The paint can now be perfectly atomized with an air pressure as low as 1.40–1.75 bar (20–25 psi) and the fluid cut down by 20 per cent or more, still maintaining the original film build. A short test on scrap material will enable the operator to correlate the air pressure and fluid flow so that he can apply a good smooth coating. Since solvents are evaporated by the heat, a smaller proportion of them reach the surface than in the case of cold spraying. Rapid cooling of the material reduces any tendency to run or sag. The paint particles hit the surface at a much lower speed, and the rebound, with loss of paint through spray fog, is reduced to a negligible minimum. This reduction of spray fog also gives the operator healthier working conditions, particularly when spraying in enclosed areas.

Time saving with heated paint

There is an alternative major advantage in spraying heated paint which sometimes appeals to the commercial vehicle refinisher, particularly when speed is essential to meet a contract. Spraying time can be almost halved simply by applying in one operation a coating twice as thick as the one normally applied. This is achieved by using the same material but without adding any thinners at all. The paint viscosity is reduced by heat to approximately the same as it is when thinners are added for cold spraying and a normal air pressure of 3.50–4.90 bar (50–70 psi) is required to atomize the heated material. As before, the rapid cooling of the material as it reaches the surface reduces any tendency to run or sag.

Some degree of paint saving can also be effected by this method by adding a small amount of fast evaporating solvent so that the spraying pressure can be lowered, but the relation between the amount of thinners added and the maximum thickness of coating obtainable depends on the material used and is best discovered by experience.

Sealers, melamines, metallic paints that have stearates in them and non-gloss paints with flatting agents can be used in a paint heater operating at a lower temperature of about 38 °C. Little or no advantage is gained with water-based materials or spirit stains.

The paint heater

A paint heating unit (Figure 158) is equally suitable for portable and stationary use. Water is pressurized to approximately 0.35–0.84 bar (5–12 psi) and recirculated in an enclosed system from a thermostatically controlled heater through a heat exchanger. Paint from a pressure feed tank or circulating system flows through tubes in the heat exchanger. The hot water, moving round the tubes in the opposite direction, transfers its heat to the paint evenly and effectively.

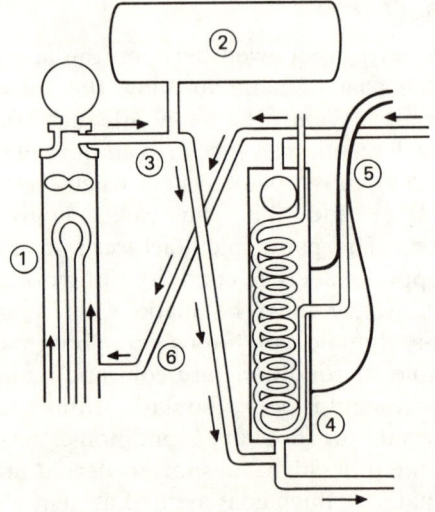

1 Hot water heater and circulating pump
2 Hot water expansion tank
3 Hot water outlet piping
4 Paint heater (heat exchanger)
5 Heated hose (to spray gun)
6 Water return piping

Figure 158 *Diagram showing action of paint heater unit*

The paint is taken from the heat exchanger to the spray gun by means of a hot water jacketed hose which is coupled into the water system, thereby ensuring that the paint is kept hot right up to the spray gun. The use of a closed hot water system is the most efficient way of heating paint and also eliminates the fire hazard and danger of polymerization that can occur in other types of paint heaters.

Electrostatic hand gun spraying

The electrostatic method of applying vehicle finishing paints has been used for many years in the automotive industry. The proven advantages in terms of uniform and consistent film build, even particle dispersion and cleanliness are equally applicable to vehicle refinishing. For metallic finishes in particular, where consistency and colour matching have always posed problems for the refinisher, the practical and economic results from electrostatic spraying are outstanding.

Principles of electrostatic spraying

The operating principle of electrostatic spraying depends on the law that unlike electrical charges attract each other. If charged paint particles are free to flow, they will follow the shortest route to the opposite charge. In practice, charged paint leaves the gun in an atomized state and earths itself on the vehicle being sprayed. Any overspray passing outside this area still has to find the shortest route to earth and consequently attracts to the edges of the work. In repelling each other the charged particles also positively assist the paint atomization process. Less operating air pressure is therefore required than for conventional spray systems.

For refinishing purposes, the electrostatic hand gun (Figure 159) will normally be supplied with paint from a remote pressure cup. A conventional air transformer directs pressure air to this container and also to the wall-mounted power unit which is connected to single-phase electrical supply. The spray gun is connected to the power unit by the multipurpose air hose which contains and shrouds the electrical leads.

Direct charging system

Paint is charged only when the gun trigger is pulled, the air pressure differential switch controlling this function being situated in the power unit remote from the gun. Charging occurs within the barrel by the direct transfer method whereby the charge is transferred to the paint column immediately before air atomization of the material takes place. This system allows a relatively low voltage value and a further advantage is that only paint particles are charged. Ionization of surrounding air and ungrounded objects is therefore avoided.

The electrostatic hand gun has no more controls than those familiar to all spray painters – a fluid pressure regulator and, positioned on the gun barrel, a spray pattern adjuster. A fluid needle adjustment is also provided. The normal connections for air and fluid supply hoses are situated on the handle and barrel respectively, but in addition the air hose contains and shrouds all electrical leads.

The aluminium spray gun body contains fluid needle and air valve components and to it is attached the insulated barrel and fluid tip assembly. All paint charging components are located within the barrel, the high voltage and

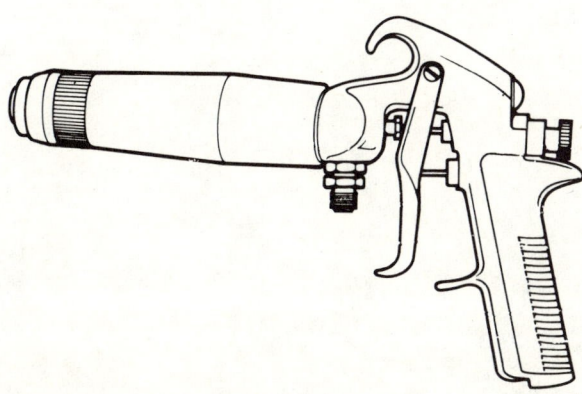

Figure 159 *Electrostatic spray gun, easily recognized by its distinctive barrel*

earth wires passing back safely through the gun body into the multipurpose air cable; there are no external electrodes.

The solid state power supply unit operates automatically and supplies a constant paint charging voltage providing that material conductivity is within the recommended limits. Specifically designed for wall mounting remote from the work area, the unit is fitted with a supply cable with earth lead and may be plugged into a single-phase standard earthed power circuit. The power unit is energized by an on/off switch and an air-operated switch in the atomizing airline controls voltage delivery to the spray gun. The air differential pressure switch is actuated only when the spray gun trigger is pulled. Paint conductivity is monitored by the microammeter built into the power supply unit. This provides a continuous check that paint and suitable solvents have been mixed correctly and that the system is operating within the optimum limits. When conductivity values are found to be outside the prescribed limits (by use of the paint conductivity test unit), polar or non-polar solvents should be added as necessary. The amount of solvent will also be determined by a viscosity test.

8 Paint materials and refinish processes

Technical terms

The following are examples of technical terms related to paintwork and recommended for general use.

Adhesion   Separation resistance between paint and base.

Anti-sag characteristic   Characteristic of paint not to run/sag while painting vertical surfaces, including thick layers.

Bleeding   Penetration of paint materials.

Chemical resistance   Resistance of paint coat against attack or infiltration of chemicals, i.e. acids, strong solutions, industrial fall-out, natural (organic) fall-out, gases.

Clouding   In metallic painting, the uneven distribution of the metallic effect, giving components/particles in the paint.

Coating   Collective term for layers made of coating materials and applied to a base.

Coating material substance   A material which is not prefabricated and provides a coating. See also *paint*.

Coat thickness   A paint layer has no compactness in a physical sense but only a thickness.

Colour   The sensory perception transmitted by the eye. A colour is characterized by colour tone, brightness and saturation (fullness).

Colour tone   Characteristic which differentiates a chromatic colour (blue, green, red, etc.) from a non-chromatic (hueless) colour (black, white).

Complete drying capacity, complete drying, hard drying   Attainment of operational hardness throughout the layer, where 'complete drying' applies more to physically and oxidatively drying (air dry) systems and 'hard drying' more to catalytically and poly-additively drying systems.

Consumption   The quantity of coating material required to provide an area of a given size under given conditions with a layer of given thickness when dry. Indication in kg/m^2 or l/m^2. See also *effective coverage*.

Corrosion protection   Sum total of measures for protecting metals, plastics and other materials against destruction by chemical and/or physical attack (aggressive media, weather, mechanical stresses).

Covering ability   Ability of a pigmented material to cover the colour or differences in colour of the base.

Drying, drying methods, drying stages   Transition of paint coat from a fluid to a solid state, caused by physical and/or chemical processes. Drying methods are, for example: air drying, heat drying (oven drying, baking or stoving) and radiation drying. Drying stages are: dust dry, tack free and thoroughly dry (fully dried)

Drying time   Period between application of paint and a drying condition determined by specified test conditions.

Effect (paint finish)   Visual appearance – under similar light conditions – of a paint finish with particles which are reflecting differently when looked at from different angles.

Effective coverage Defined as the area of a surface attained with the quantity of a coating material for one layer of certain thickness. Data in m^2/l or m^2/kg. See also *consumption*.

Filling capacity Capacity of a paint to equalize the unevenness of the base, which provides the optical effect of a particularly well painted surface (fullness).

Flow structure The general characteristic of paint in its liquid state to spread out smoothly on the surface being applied to.

Gloss Visual appearance of various reflection of light on a paint finish – high gloss, satin gloss, etc.

Gloss retention Characteristic of a paint finish to resist dulling, that is, of a paint finish dulling by usage, external influences, etc.

Hardness Mechanical property of paint coat. Expressed by resistance of paint to mechanical effects (e.g. pressure, friction, scratching).

Hazing Clouding of a paint surface not caused by external influences which shows up during or after drying.

Paint substance Substance which provides a paint coat with specific characteristics. Collective terms can be used with word combinations, such as primers, stoving enamels, dipping enamels, clear coating, etc.

Paint, paint material From liquid to pasty, also powdery coating material: applied on a base by means of pertinent methods. Provides a solid paint coat after a physical and/or chemical drying process.

Paint coat, paint film Paint layer on top of a primer coat. Paint film is another term defining the formation of a cohesive layer.

Paint foundation Collective term for basic materials such as steel, aluminium, plastics which serve as the base for the respective paint finish.

Penetration (staining) (a) Appearance of stain from the undercoat or an existing paint coat into paint layers above. (b) Appearance of paint stain on back of base (e.g. paper).

Pigment separation Partial separation of binder and pigment or of pigments from each other.

Polishable Characteristic of a paint for polishing (buffing) with polishing equipment (buffing pad, buffing wheel) and polishing compounds (paste, wax, polish) to obtain a high gloss.

Primer, primer material Paint material suitable for priming.

Prime coat Prime coat (primer) as a paint layer suitable to serve as a connection between the base and paint coats.

Rub resistance Resistance of paint finish to light rubbing.

Runs, sagging Characteristic/tendency of paint to form runs when painting vertical surfaces. See also *anti-sag characteristic*.

Sanding ability of paint coats during wet or dry sanding: (a) for smoothing and/or uniform roughing as a preparation for an additional paint coat; (b) as a characteristic, especially for priming, without faults (mechanical resistance, etc.) for preparing with sanding equipment.

Scratch resistance Resistance of a paint coat to scratching influences, depending on hardness and drying condition.

Sedimentation Formation of sediment of pigments caused by the joining of particles which are specifically heavier than the binder solution and will therefore settle to the bottom.

Separation Visible separation of pigments in paint during storage or on paint coat during liquid state.

Swelling (a) Increase in volume by absorbing liquids, vapours or gases into paint film. (b) 'Wetting interference' at the edges of a sanded area.

Weather resistance Characteristic of a paint coat, defined as its resistance against changes: (a) exposure to the elements through effect of weather at the test site; (b) quick test by the effect of simulated exposure to weathering.

Wetting Ability of paint to adequately flow, without interference.

Wetting interference The appearance of flaws, i.e. fish-eye in a paint film.

Yellowing Change in colour, especially of bright colours, caused by high thermal stresses or strong light effects.

Paint characteristics

The term 'paint' can be used to describe any fluid that can be applied to a surface and will dry to form a continuous hard film. Refinish paints vary in their properties and uses but they all have three basic components in common: pigment, binder or vehicle and solvent or thinner.

Pigment

Pigments used in the manufacture of paint are finely ground powders. These may be derived from naturally occurring minerals or they may be synthetic dyestuffs. Their properties are very important because they give the paint its hiding power (opacity) and colour and help to determine its durability. The actual pigmentation of a paint depends upon its function. In primer–fillers they are chosen to give good build and easy flatting. In finishes they are selected to give a lasting decorative effect.

Binder or vehicle

This supplies the film-forming properties to the paint, binding the particles together and providing the adhesion to the substrate.

Solvent or thinner

This makes the pigment/binder mixture fluid and workable during paint manufacture. It also reduces the paint to the correct consistency for application by spray gun, brush or knife. The solvent mix is volatile; it evaporates once the paint has been applied, leaving the pigment and binder to form the hardened paint film.

The drying process

The full drying process is not the same for all paints; broadly, paints can be classified into four groups:

1. In those paints generally known as lacquers, drying occurs purely as a result of solvent evaporation and there is no chemical change involved. The characteristics of the group are of a high solvent content and rapid initial drying. The group includes:

    Cellulose- and acrylic-based car repair paints.
    Cellulose-based quick-drying primers, primer–fillers, sealers.
    Cellulose-based stoppers.

2. This group contains oil- and synthetic-resin-based paints. In these, initial drying occurs by evaporation of solvent, but final hardening is due to chemical changes in the paint vehicle caused by the uptake of oxygen. With this type of material there is always a critical period during which wrinkling or lifting will occur on recoating, owing to the chemical changes which occur in the vehicle when this is gradually becoming less soluble in its own solvent. The group is characterized by a high solids content, and a slower surface dry time than with lacquers. It includes:

    Synthetic quick-drying spraying finishes.
    Synthetic coach enamels.
    Oil- and synthetic-resin-based primers, primer–surfacers, stoppers.

3. Paints in the third group will not air dry but to fully harden need a baking period at a

temperature of 80 °C upwards. Solvent evaporation causes initial drying to take place. But final through hardening is dependent upon a chemical reaction between two components in the paint vehicle, and this reaction can only take place in a time period of about 30 minutes after the stoving temperature has reached 80 °C. This group is characterized by a low solvent content and the need to install expensive stoving equipment. It includes low-bake enamels.

4 The fourth group consists of two-pack paints. These materials will not harden until both component parts have been mixed together when a chemical reaction takes place causing the paint vehicle to harden. One disadvantage is the limited pot life or working time once the two component parts have been mixed. The group is characterized by a low solvent or even solvent-free content and a high vehicle content. Typical materials are:

Two-pack polyurethane finishes.
Two-pack epoxy primers and finishes.
Polyester stoppers and spraying fillers with peroxide catalysts.

Undercoats

Undercoat is a term used broadly to describe the coatings which can provide a base for the final colour coats. It includes primers, primer–surfacers, fillers, stoppers and sealers, all of which can belong to any of the four groups.

Primers

A primer is an undercoat designed to ensure that the paint system adheres well to the substrate, initially and in service. It is generally used over bare metal, although it can also be used over old finishes. As its main function is to establish a secure base for the paint system as a whole, a primer is most effective if applied in a relatively thin film. Primers are not formulated to fill scratches or to be flatted. A very light denib is all the rubbing down that should be carried out.

Etching primers are a special type of primer containing acid, to give metal etch. In addition to performing the normal functions of a primer, they also act as a metal pretreatment by preventing the spread of corrosion if the paint film becomes damaged in service. They are most effective when the primer base and acid activator are supplied separately and mixed together just before use.

Primer–surfacers

A primer–surfacer has a dual function – to provide good adhesion and to fill minor scratches. When dry, the primer–surfacer is flatted to level surface imperfections and provide a smooth surface, so ensuring a uniform gloss in the colour coats which follow.

For better filling and easy flatting, primer–surfacers contain a much larger amount of pigment than pure primers. This high pigment content leads to poor water resistance if a primer–surfacer is applied in a thin coat on bare metal like a primer. Thus it is important to apply a minimum of two full coats.

Fillers and primer–fillers

Fillers are heavily pigmented materials; they are designed to fill deeper scratches and metal irregularities than are primer–surfacers. It is generally possible to apply a greater thickness in one spray coat of filler than in one spray coat of application primer–surfacer. Because of this, the drying or 'flashing' time between coats should be carefully controlled and the number of coats applied in one day restricted, otherwise the time required for final hardening will be considerably lengthened.

The same restrictions may not always apply to products which harden due to chemical reaction of two vehicle components mixed just before use. For example, polyester spraying fillers which harden with peroxide catalysts can be used to give high build very quickly.

Fillers can be formulated for application by spray, brush or knife. Knifing fillers will give the best filling, but they call for considerable skill in application. Spray fillers can be applied easily but they will follow the contour of the surface being painted and must then be rubbed to obtain a level surface. The brushing filler will fill better than the spray filler but considerable flatting will be necessary to eliminate the brush marks. Some products which come into the filler category, but which can be applied to metal, are called primer–fillers.

Stoppers

A stopper is a heavily pigmented undercoat applied by knife, and used to fill deep scratches or indentations in the surface. It is intended for application over small areas only. For knife filling over large areas, a knifing filler should be used.

Sealers

A sealer is used to improve the uniformity and gloss of the final colour coat, to fill flatting scratches, and to improve colour hold-out. It can also be used as a ground coat to aid the final colour match. Sealers should not be expected to act as 'isolators', to prevent wrinkling of an old paint film under the action of strong lacquer solvents, or to stop bleeding through of a previous colour coat. Special products called 'bleeding inhibitor sealers' are available to stop bleeding. A point to note about these is that they give a poorer hold-out under colour coats than the standard sealers.

Ground coats

A ground coat is used to obtain a surface of uniform colour over which finishing colour can be applied to give a solid colour effect with a minimum number of coats. It is particularly useful under certain yellows and reds which have rather lower opacity. Flatting is not recommended on ground coats as it may destroy the colour uniformity.

Finishes

Nitrocellulose lacquers

These lacquers are a blend of nitrocellulose and synthetic resins. The nitrocellulose content imparts rapid drying characteristics. The synthetic resin content gives the high build, gloss-from-the-gun colour and gloss retention properties. Nitrocellulose lacquers dry by evaporation of solvent only. As the initial rate of solvent evaporation is extremely rapid, so is the speed of surface dry. However, some solvent is retained for a longer period. For full dry and hardness, a 16 hour drying period is necessary.

When properly applied a nitrocellulose lacquer will dry to a level film with a high gloss. If necessary polishing can be carried out to remove slight imperfections in the film, and to achieve a higher gloss level.

As the final appearance of the paint film is dependent on the use of a good-quality thinner, it is essential to use the one recommended by the paint manufacturer.

Acrylic lacquers

Acrylic lacquers are a blend of a resin from the acrylics family, such as methyl methacrylate (or 'Perspex') and a synthetic plasticizing resin. The characteristics of acrylic lacquers are very rapid surface drying, excellent gloss and colour retention. Drying is by solvent evaporation only. This gives the rapid surface drying, but it means that for full through hardness a 16 hour drying period is necessary. Acrylic lacquers respond readily to hand or machine polishing, using rubbing or polishing compounds. Usually some degree of polishing is necessary, depending on the gloss level required.

Properly applied, an acrylic lacquer system can give excellent results. For durability the undercoats must be those recommended for acrylics. The manufacturer's instructions should always be followed. If products from different manufacturers are mixed, failure of the paint film may occur by cracking or crazing.

Air-dry synthetic enamels

Refinish air-dry enamels are based on synthetic resins modified during manufacture by drying oils such as linseed or soya bean. All enamels dry in two stages. First, the solvents or thinners evaporate from the paint film, which may then be handled with care. In the second stage the paint film hardens by the absorption of oxygen from the atmosphere. After overnight drying it will be hard enough for the motor vehicle to go into service.

The solids' content of enamels is very high. Two coats will give a high-gloss film with excellent scratch-filling properties. The solvents used in synthetic finishes are comparatively weak so there is no danger of lifting or crazing old paint films.

Low-bake enamels

In low-bake enamels, the paint vehicle consists of a blend of synthetic resins which will harden only when the paint film reaches a temperature of 80 °C. Stoving equipment must be installed to use this type of paint which is similar in character to the original stoved factory finish on most mass production cars. Application solids of low-bake enamels are very high. As with air-drying synthetics, two coats will give a high gloss film with excellent scratch-filling properties. Solvent mixtures are comparatively weak, and there is little danger of old paint films either lifting or crazing. After spraying, 5 to 10 minutes should be allowed for the paint to flash off. Then the motor vehicle should be placed in the convection oven for 40 minutes at an air temperature of 90–95 °C. This air temperature is necessary to ensure that the lower panels of the vehicle reach the minimum baking temperature of 80 °C.

Lacquer finishes which dry by the evaporation of solvent only can also be force dried or low baked. They will reach full hardness at temperatures some 10 °C lower than true low-bake materials. An extended flash-off time is generally required to avoid popping. Air-drying synthetic enamels can be converted into low-bake enamels by thinning to spraying viscosity with special resin/thinner solutions generally known as hardeners or hardener/thinners, and thinner is necessary to reduce the enamel to spraying viscosity.

Full low-bake temperatures are not necessary to harden converted enamels but panel temperatures of 60 °C and upwards are recommended where high production rates are required. Typical stoving schedules would be 40 minutes at 60 °C or 15 minutes at 80 °C metal temperature.

While hardness from the oven will not be quite to the high standard obtained with a true low-bake enamel, the motor vehicle may safely be put into service after stoving and the paint film will then reach full hardness after overnight drying.

Two-pack acrylic enamels

In high-bake acrylic enamels, which are used extensively by manufacturers in the original finishing of motor cars, the paint vehicle consists of a blend of acrylic and melamine resins which will only harden at a rate to meet production requirements when the paint film reaches a temperature of 125 °C. Similar acrylic resins are suitable paint vehicles for refinish enamels when they are mixed just before use with a hardener based on an isocyanate resin. They will harden at normal shop temperatures and the hardening process can be accelerated by force drying at temperatures up to 80 °C.

The two-pack acrylic refinish enamels have high application solids, and two coats will give a high gloss film with excellent scratch-filling properties. They will cure at a faster rate than other refinish paints at any drying temperature from 20 °C to 80 °C and final hardness is equivalent to that of the high-bake acrylic production finish. Recoat can be carried out and petrol resistance is achieved after drying for 6 hours at 20 °C or at a typical force-dry schedule of 30 minutes at a metal temperature of 60 °C. After drying overnight at 20 °C or force drying

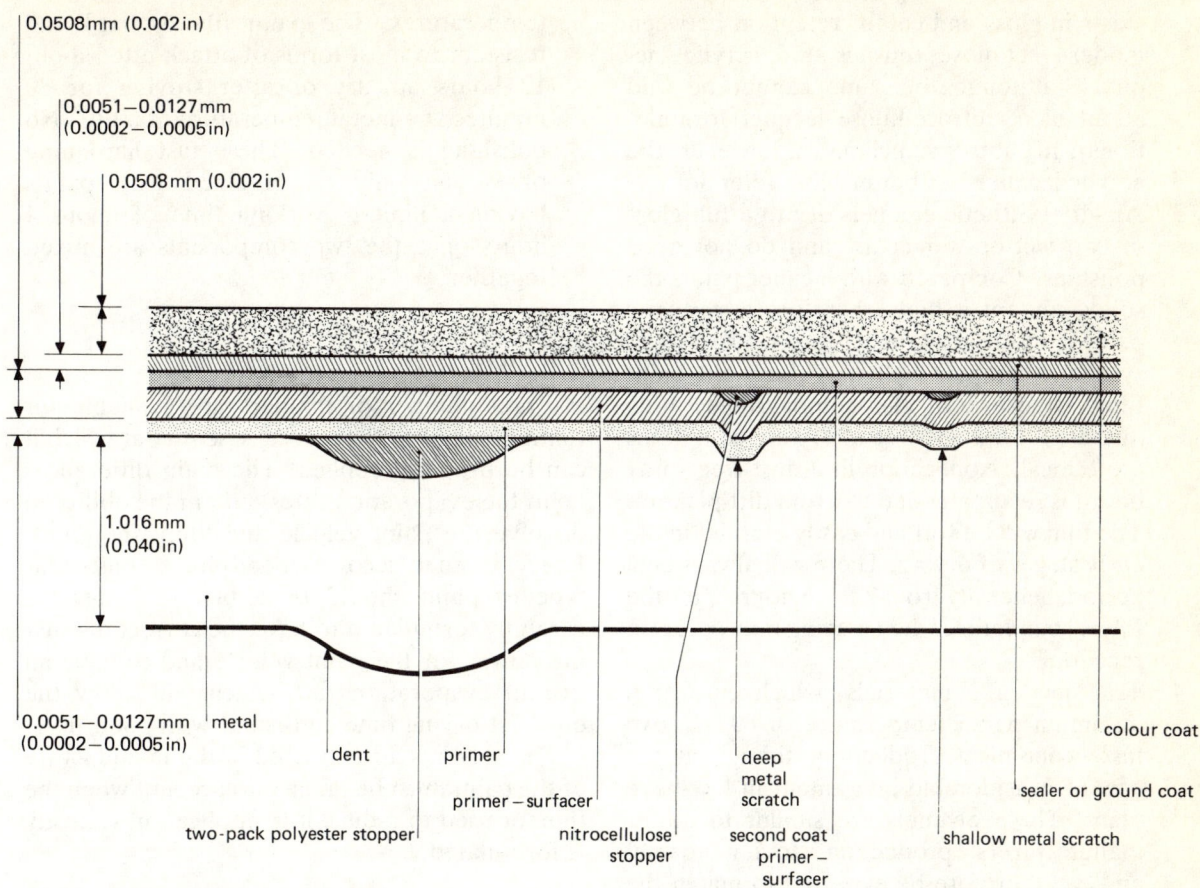

Figure 160 *Structure of a paint finish – location and approximate film thicknesses of the various refinishing materials*

either as above or at any other appropriate schedule, e.g. 15 minutes at 80 °C, the film can be flatted and polished to remove dirt inclusions or even quite severe sags.

The pot life or working time of these paints will generally be in the region of 4 to 8 hours. After this time all mixed paints should be discarded.

Figure 160 gives the location and approximate film thicknesses of various refinishing materials.

Difference between the various refinishing materials

While the differences in application characteristics of lacquers and enamels have narrowed, the introduction of refinish materials such as acrylics and low-bake enamels has made the choice of the refinish system wider than ever. The main differences in finishing paints are:

1. Lacquers surface dry rapidly and dirt in the work area is not a problem. The solid content is comparatively low and high paint usage is necessary to obtain full gloss and good scratch filling. Lacquer spraying thinners are strong, and may cause old paint films to wrinkle, so great care must be taken when repairing the older motor vehicle.
2. Acrylic lacquers surface dry slightly faster than 'glossy-from-the-gun' nitrocellulose lacquers. In temperate climates, little difference

exists in gloss and colour retention between modern cellulose finishes and acrylic lacquers (although the same cannot be said about older nitrocellulose lacquer formulations). In subtropical climates, however, the acrylic lacquer has better gloss retention.

3. Air-dry synthetic enamels dry to a full gloss in two wet-on-wet coats, and do not need polishing. Compared with lacquer paints the solids content is high; therefore a synthetic paint is more economical to use. The weak synthetic thinners will not craze old paint films. Because of the comparatively long dust-free time involved, drying conditions are critical. Application in a dust-free spray booth is recommended to avoid dirt pick-up. The film will be soft and easily marked in the early stages of drying. There will always be a period, generally from 2 to 16 hours after the job is completed, when crazing may occur on recoating.

4. True low-bake materials, which cure at a minimum panel temperature of 80 °C, give fast economical production, though at the cost of considerable investment in low-bake plant. These enamels are similar to motor manufacturers' production stoving enamel and can therefore be expected to match the production finish in every way. Paint film is hard and resistant to most forms of attack, straight from the oven. No polishing is needed.

5. When air-dry synthetic enamel is used with a hardener or thinner, the production advantages gained from using a low-bake oven are obtained without the need for the critical stoving schedules which true low-bake enamels require. However, hardness from the oven is not quite so good; the paint film will be more easily damaged in the trim fitting operation and some lifting may occur if areas with a high film thickness are recoated. Again, no polishing is needed.

6. Two-pack finishes will give more rapid hardening than other high solids refinish enamels in air dry and at low stoving temperatures. The paint film is hard and resistant to most forms of attack after about 12 hours air dry or after stoving for 30 minutes at a metal temperature of 60 °C. No polishing is needed. These fast hardening times can only be obtained with paints having a limited working time of about 4 hours once the two components are mixed together.

Thinners

A great number of thinners are available for thinning paint to the correct viscosity at which it can be properly applied. The main differences lie in the evaporation rates, and in the ability to dissolve the paint vehicle and 'thin' the paint. Every thinner recommended for a particular type of paint should be a blend of solvents carefully formulated to have the correct dissolving power for the paint vehicle and to have an overall evaporation rate which will allow the quickest drying time consistent with good flow.

The solvents already used in the manufacture of the paint must be taken into account when the thinner used to reduce it to application viscosity is formulated.

Metallics

Metallic finishes for cars have been described by many names, such as metallic, polychromatic, metallescent, Starfire, Starmist and opalescent. Of all these names, metallic is probably the best since it tells the refinisher it contains metallic particles or pigment. Since metallics were introduced there has been a constant increase in their use on new motor cars, which, therefore, constitute a large proportion of the cars requiring repairs to metallic finishes. Metallic finishes cause more problems for the refinisher in the form of application and colour matching than any other finish in the refinishing world. The chief reason is that one is not only matching colour but appearance, which is affected by application.

Formulation

A straight colour contains only finely dispersed pigments of colour matter, usually in sufficient quantity to give complete opacity within the normal thickness of the paint film, 0.0508 mm (0.002 in). Light is reflected from the paint surface in a normal manner (as from a mirror) and usually does not penetrate the paint film beyond 0.0254 mm (0.001 in) (see Figure 161).

On the other hand a metallic paint usually contains less coloured pigment than a straight colour, and without the presence of aluminium particles the light would penetrate right through to the undercoat; in other words the opacity or covering power would be inadequate without these aluminium particles. The aluminium particles reflect the light rays before they reach the primer, but the light will be reflected at different angles (see Figure 162). For this reason, metallic colours have to be matched for both face and bottom tones.

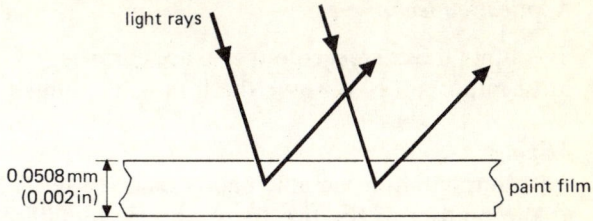

Figure 161 *Light reflection from straight colour paint film*

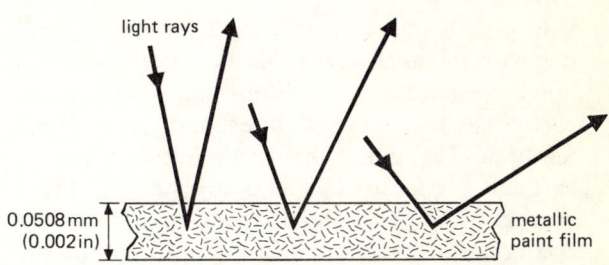

Figure 162 *Light reflection from metallic paint film*

Wet application

Because the paint is very wet it takes longer to dry and the aluminium flakes have time to settle in the layer of paint. The light which penetrates the paint is reflected back from the aluminium flakes through the translucent part of the paint film (Figure 163). The aluminium flakes also tend to assume a more upright position in the paint film. This gives the colour a deeper top tone and a paler bottom tone.

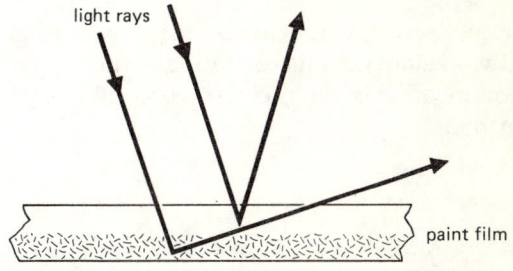

Figure 163 *Wet application*

Dry application

When a metallic paint is applied dry, the aluminium flakes have no time to settle as the paint film flashes off immediately. The light is reflected from the aluminium as in a wet application but, as the aluminium flakes have not had time to settle, one does not see as much of the translucent part of the paint; therefore a lighter, greyer top tone and deeper bottom tone is produced (Figure 164).

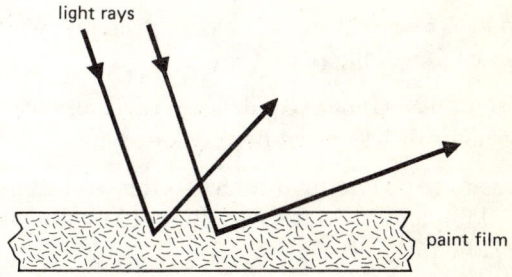

Figure 164 *Dry application*

Application technique

To obtain the correct colour and appearance it is most important to observe the following points:

Mixing

The aluminium in metallic paints tends to settle at the bottom of the tin; therefore the metallic colours need to be stirred more thoroughly than straight colours.

Thinning

The ideal way is to empty the tin out and wash out with thinners so that all the paint is thinned and a clean tin is left. It is also important to add the correct type of thinner and the correct amount. The paint should then be measured through a viscosity cup and adjusted until it is correct. This will help to achieve the correct colour and appearance. It is important to stir thinned paint very thoroughly. If it is left to stand before being used, the aluminium settles very rapidly.

Gun set-up

The gun set-up will depend on the type of gun and workshop conditions. Often a gun set-up is recommended in the product technical information book.

Terms associated with metallic colours

In addition to the terms common to solid colour there are several terms used to describe various colour aspects of metallic paints.

Tone Tone relates to colour strength in the case of a metallic paint film.

Undertone Undertone denotes the presence of a weaker or less prominent colour.

Mass tone Mass tone refers to the overall hue or dominant colour, i.e. the colour of the metallic finish.

Top tone (face tone, direct view) Top tone describes the tone of a metallic when viewed perpendicular to the surface, i.e. as viewed directly into the finish.

Bottom tone (side tone, flip tone, oblique view) Bottom tone relates to the variation when a metallic is viewed at an oblique angle, i.e. from nearly parallel to the surface.

Flip Flip is often used to describe the colour contrast between top and bottom tones.

Metallic appearance/metallic effect The term appearance when used to describe a metallic paint finish denotes the prominence (brilliance) of the aluminium flakes.

Factors affecting colour of metallic finishes

Colour tone variations are often confused with colour chroma (i.e. 'shade') variation. Although these are interrelated, colour shade variation and many matching problems arise purely as a result of neglecting a few simple principles of mixing, thinning and application.

To obtain the correct colour and appearance it is most important with all metallic paints to observe the paint manufacturer's recommendations. Correct preparation and methods will also depend on the type of metallic structure which may be either single-coat or clear-coat metallic technology.

The colour of metallics can be influenced by many different factors. After doing the common-sense things, i.e. stirring the paint, cleaning the equipment, etc. the colour can be affected by:

1. Working conditions
2. Spray gun adjustment
3. Thinners
4. Spraying technique

Table 6 outlines the variables affecting the colour of metallic finishes.

Metallic technology and techniques

Single-coat metallic colours

'Single coat' implies the conventional type of

Table 6 *Variables affecting colour of metallic finishes*

			Material		
Variable	To lighten colour	To deepen colour	Cellulose	Acrylic lacquer	Synthetic
Spray booth conditions					
Temperature	Raise	Lower	*	***	*
Humidity	Low	High	*	**	**
Air movement	Increase	Decrease	**	**	***
Spray gun					
Fluid tip	Small	Large	*	*	***
Needle control	Close up	Open out	*	*	***
Air cap	Use high air consumption cap	Use low air consumption cap	***	***	***
Fan width	Widen	Narrow	**	**	**
Air pressure	Raise	Lower	**	**	**
Thinning					
Type of thinner	Use fast thinner	Use slow thinner	**	***	**
Amount of thinner	Lower viscosity	Raise viscosity	***	***	**
Retarder thinner	Do not use	Add 10% max.	**	**	**
Spraying technique					
Gun distance from surface	Increase	Decrease	**	**	**
Gun speed	Increase	Decrease	*	*	*
Flash-off time between coats	Lengthen	Shorten	*	*	*

Note: Number of asterisks indicates degree of importance.

metallic finishes with uniform paint film structure and utilizing aluminium flakes (or particles) as inherent pigment. The term is not intended to imply that this type of metallic paint can be applied in a single coat of paint.

Dusting technique (dust coat)
The dusting technique is a spray application technique performed to lighten a metallic tone or to create a coarser metallic appearance. The method involves applying a final coat in the form of semi-dry particles to the *wet* metallic film. This is achieved by increasing the distance between the spray gun and the surface to approximately 400–600 mm (16–24 in) and applying the paint lightly in the form of a fog.

Cross-spraying (cross-coats)
Cross-spraying is a technique that may be used during application to ensure even film thickness

and build, to eliminate shadiness or a shady appearance of a metallic paint film. This method involves spraying the paint in light coats, first a normal coat using horizontal gun strokes and then a coat using the spray gun in a vertical direction. Consecutive coats are applied in pairs and wet on wet (effectively as a double coat).

Clear-coat metallics

An additional metallic paint structure recently introduced to the refinisher as a two-coat system called clear-coat metallic. Clear-coat metallic or more explicitly base-coat and clear-coat paint films comprise an integral two-layer structure: the pigmented base layer with conventional or special brilliant aluminium flakes is overlaid with a final transparent layer. The base layer or base-coat provides the colour reflective properties, the opacity and the metallic appearance of the complete paint film structure; it will flash off and dry to a semi-matt condition. The second coat or clear-coat provides the film with a smooth high-gloss protective coat for the underlying colour and aluminium flakes in the base-coat.

The terms applicable to conventional single-coat metallics relate also to clear-coat metallic colour structure.

Refinish processes and operational sequences

In order to match as far as is possible, the high standards of protection and finish provided by the vehicle manufacturer, the use of top-quality refinishing materials obtainable from approved suppliers is imperative. Typical refinish processes and their sequences of operation are as follows.

Small-blemish panel beating

By using the various hammers, picks and dollies contained in the Ding Kit (Figure 165), many small blemishes in the sheet metal body can be removed without damage to the paintwork and the consequent necessity to refinish the repair area.

Polish

Metallic finishes must be treated with discretion since local colour of the paint film can be affected by robust polishing.

1 Wash and leather dry area(s) to be polished to remove any surface road dirt or grit.
2 Mask off any air intakes or outlets on the body in repair area to avoid contaminating the ventilating system.
3 Brush or dab rubbing compound on to areas to be polished. Before switching on machine, wipe mop across the work area in order (a) to distribute the polishing compound evenly, and (b) to impregnate the mop with compound.
4 Use light overlapping parallel strokes to polish the defect and entire panel(s) on which the defect is apparent.

In some instances defects may not be entirely removed by polishing compound alone; where this occurs, 600 grade abrasive sanding paper can be used to assist removal of defects by hand sanding, using water or white spirit as a lubricant. If abrasive paper is used, sanding debris must be wiped or blown from the area to be polished before repeating operations 3 and 4.

5 When the defect has been removed, wipe the area with a clean cloth moistened with naphtha and lightly polish the entire panel (using a clean mop) before removing any masking applied.

Note: a sprinkle of clean water over the panel surface will produce maximum gloss.

Brush touch-up

Where a chip or scratch is minor in nature and confined to a panel edge, or is low down on the body, a satisfactory repair can usually be made

Paint materials and refinish processes 151

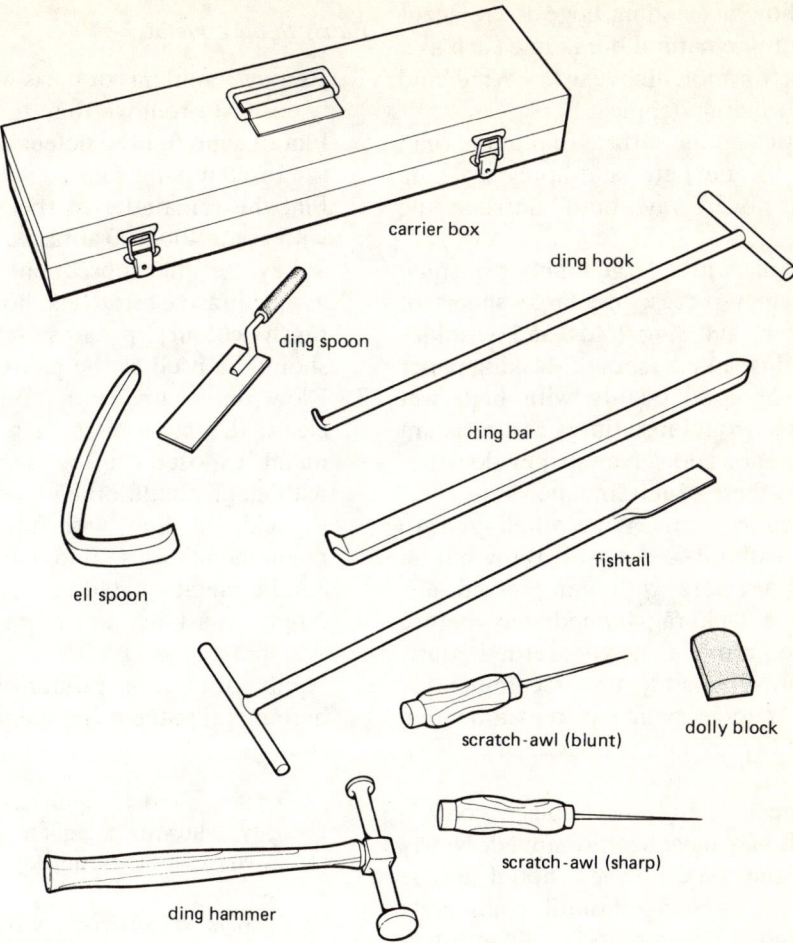

Figure 165 *Ding kit*

by careful application of suitable air-drying body colour applied by brush.

1. Thoroughly stir the paint to ensure an accurate colour match and dispersion of constituents.
2. Pour sufficient paint into a small clean container, adding sufficient thinner to bring the material to a brushing consistency.
3. Clean area to be painted.
4. Carefully apply paint to the defect with a small brush. Allow to dry, examine the repair and – if necessary – apply a second coat to fill the defect.
5. When hard, blend repair by lightly polishing the repair and adjacent areas by hand, using fine grit compound.

Colour coat

1. Remove any emblems which are normally easier to remove than to mask.
2. Flat or sand out any defects to obtain a level surface and provide a key essential for intercoat adhesion. Prepare the original

finish to allow a blending edge to adjacent panels, where no natural break line such as a coach joint or moulding exists. Wipe and blow away sanding debris.
3. Thoroughly clean the surface to remove road dirt, wax, silicones, etc. and apply masking to prevent overspray onto surrounding panel(s).
4. Avoid masking with a large number of small tape and paper pieces; use large sheets of brown paper, avoiding folds and wrinkles which trap dust and dryspray. Masking paper should be of good quality with high wet strength, free from loose fibres and resistant to solvent penetration. Newspaper does not meet any of these requirements.
5. Apply the colour coats. Wipe off all areas to be painted with oil-free petrol. Blow out all seams and crevices with compressed air. Wipe with a tack rag immediately before painting to remove newly settled dust. Thoroughly stir paint, use recommended thinner to reduce paint to recommended viscosity, fill spray gun container and check spray pattern.
6. Remove the masking and replace components which may have been removed. Newly applied lacquer-type finishes should not be polished or compounded until completely hard. Compounding too soon will result in loss of gloss. Solid wax polish must never be used until the finish is three to four weeks old; wax will migrate into the finish, causing softness, loss of gloss and possibly water spotting. Hand polishing should be carried out with a soft open-mesh cloth, in straight lines and following the styling lines of the repaired area. Rubbing compounds may be regarded as flatting papers; coarser grades cut faster but produce deeper scratches. *Do not* carry out heavy polishing operations on thin final coats or metallic finishes. Alternative materials such as metal polish should be avoided as these contain strong solvents, alkalis and ungraded grit particles which may stain a light-coloured film or scour and scratch a new finish.

Patch to bare metal

1. Remove emblems or parts which are normally easier to remove than to mask.
2. Flat or sand out the defect and 'feather edge' the broken paint film edge to existing finish. Flat the remainder of the affected panel(s) with grade 400–600 abrasive paper to provide a key for the subsequent paint films. To avoid abrasive scratches showing through the finish colour, progressively finer grades should be used as the paint is built up.
3. Blow out seams and mouldings etc. with clean, dry compressed air. Treat any bare metal exposed during sanding operations with metal conditioning fluid, e.g. phosphoric acid, in line with the manufacturer's recommendations, and neutralize. Do not handle metal treated in this manner.
4. Apply masking as required to prevent overspray.
5. Apply primer or primer–surfacer coats to bare metal patch using a uniform wet, single coat.

Note: Single coats sprayed at recommended viscosity, allowing adequate flash-off time between coats, will dry quickly and give effective filling.
Thick coats will surface dry but trap the solvent and remain soft beneath the surface so that material will sink and follow imperfections rather than fill them

6. Apply nitrocellulose-based stopper and recoat with primer after flatting. Ensure that each application is completely hard before proceeding.
7. If the primer patch is large, it is advisable to apply a guide coat, before wet sanding. This ensures that the minimum amount of surface is removed and also highlights hollows or imperfections (guide coat colour remains in these locations).
8. Using grade 400–600 abrasive paper, flat down the primer coat to obtain a smooth surface and blend in the edge of the patch.
9. Apply the colour coats.

Strip to bare metal

1. Remove badges, emblems and mouldings, etc. from panel(s) to be stripped.

    (a) Power sanding is suitable for removing old finishes from small flat areas or large-radius curves. Remove paint with 24 or 36 open-coat abrasive discs, followed by 50 grit discs and 100–120 grit discs to reduce scratch level.

    (b) Chemical paint removers can be used for stripping large areas but adjacent non-defective panels must be protected by masking foil which is resistant to chemical paint remover. Scoring the surface to be stripped will speed penetration of the fluid, but some undercoats may also require sanding for complete removal. Scour off residues with cleaning solvent and steel wool, followed by an immediate wipe with clean dry rag.

2. Chemically clean the bare metal surface with a metal conditioning fluid of the phosphoric acid type to remove rust and provide a key for subsequent paint coats.
3. Apply masking as required to prevent overspray.
4. Apply the primer and/or primer–filler coats as necessary. Etch primer is necessary on aluminium and has advantages on steel, promoting good adhesion, corrosion and blister resistance. It is particularly suitable for sills and wheel arch areas. It is water sensitive, however, and must be covered with primer or primer surfacer as soon as possible after application.
5. Apply nitrocellulose-based stopper if required and recoat with primer after further flatting.
6. Apply a guide coat of paint to the primer covered panels when dry.
7. Wet sand the primer coats, using grade 400–600 abrasive paper. The flatting or sanding operation is carried out to obtain a level surface and provide a tooth or key essential for good intercoat adhesion.
8. Apply the colour coats.

Oxalic acid wash

If a vehicle has been stored or used for lengthy periods in an industrial area, ash, grit, etc. (sometimes containing iron particles) from factory chimneys can become embedded in the paint film through chemical action.

If the contamination is very slight it can be removed by a light polishing operation, but generally this method should not be used since the disturbed 'fall-out' particles are frequently extremely hard and could be picked up by the polishing cloth or mop and cause surface scratching. A simple and effective cure is to use a chemical wash.

Prepare a mixture of oxalic acid crystals in water in the proportions of 12 to 25 g/l (2 to 4 oz/gal) of water and to this mix add a few drops of liquid detergent.

Note: Oxalic acid is poison and must be handled with care. It must be clearly identified as a poison, and stored under secure conditions. Skin contact must be avoided by use of rubber gloves and any splashes must be washed off well with water. The eyes should be protected by goggles, particularly during mixing or dissolving. Any splashes in the eye must be irrigated with water or saline (salt) solution for 15 minutes and medical attention sought at once.

It is essential that the operation is carried out in a location which will ensure that the drainage takes place into a foul water sewer, and not to a surface water drainage system. The local water authority should be informed of an intention to use this process, to avoid contravening pollution or drainage laws.

The solution will not react if it is allowed to dry. For this reason it is not advisable to carry out this procedure in strong sunshine.

1. Thoroughly clean the car with cold water.
2. Carefully wipe all surfaces to be treated with a wax/grease removing solvent. Remove all bituminous stains with white spirit.
3. Apply the solution to the affected areas (normally horizontal surfaces only) using an 80–100 mm (3–4 in) soft brush. Care should be taken to prevent the solution running behind mouldings or badges.

4 Keep the film wet and active by reapplying the solution and allow it to remain for approximately 15–20 minutes.
5 Thoroughly wash and flood off the solution by hose. This is essential to ensure that all traces of the solution are removed from the painted surfaces and from behind mouldings, etc.
6 Closely examine the affected surfaces – usually one treatment in this manner will be sufficient to remove all traces of industrial fall-out. The car can then be dried with a leather.

Spot repairs

Spot repairs are recommended only where a complete repair of the damaged panel would be uneconomical owing to its size or the amount of masking involved. This type of repair is normally carried out using enamel-type finishes and not with synthetic materials, because of difficulty in polishing the dry edge and obtaining colour blend. When spot repairs are made, a narrow spray pattern and reduction in fluid delivery and air pressure will minimize overspray.

1 Carefully feather edge the damaged area, finishing off with paper not coarser than 400 grade used wet.
2 Build up using a lacquer-type primer–surfacer working from the centre outwards allowing each coat to overlap the previous one slightly.
3 Air dry, or force dry, primer–surfacer thoroughly; this is essential if sinkage and contour mapping are to be avoided.
4 Wet flat using paper not coarser than 360 grade. At this stage primer–surfacer overspray must be removed from surrounding areas. Feather edge of surfacer may be improved at the same time by burnishing with rubbing compound.
5 Spirit wipe (particularly when rubbing compound has been used).
6 Tack rag the surface and apply colour coats. Work from the centre outwards. Keep colour spot well inside flatted or compounded area to ensure an adequate bond to the original finish.
7 For final coat, use mixture of 9 parts thinner to 1 part colour and apply lightly to the dry spray at the edge of the spot as soon as possible after the last coat of full colour has been applied.
8 After adequate drying, polish the repair using rubbing compound.

Special processes

Dewaxing

New vehicles coated with wax to prevent deterioration from atmospheric contaminants during storage or transit must be properly dewaxed using the correct procedure. An incorrect process may damage the paint film or create a defect which may only reveal itself at a later date. There are two acceptable methods of dewaxing:

1 Steam cleaning.
2 Dewaxing fluid.

Paint must not be exposed to excessive application of either steam or solvent. Use of incorrect solvents may cause damage to trim, rubbers or weather seals.

Protective clothing should be worn. Use either method only in a well-ventilated area and not in the proximity of operation areas.

Steam cleaning

1 Use an approved steam cleaner correctly adjusted to deliver wet steam consisting of 75 per cent hot water at a pressure of 5.25 bar (75 psi).
2 Fill the detergent tank with water-soluble dewaxing fluid.
3 The steam delivery jet must not be allowed closer than about 75 mm (3 in) to the paint surface while slowly working over the panels.
4 Wash off all sludge with a jet of hot water.
5 Dry off completely with a chamois leather.
6 Polish out smears.

Dewaxing fluid

1. Apply solvent to waxed surface with a sponge, by spray or a cloth.
2. Allow to react for 5 minutes or the recommended period. The solvent must not be left in contact with the paint surface for longer than necessary.
3. Wipe off softened wax with clean cloths.
4. Wash off thoroughly with clean water or a solution of water-soluble cleaner and water.
5. Rinse with cold water.
6. Dry off with a chamois leather.
7. Polish out smears with a clean soft cloth.

Certain conditions can result in wax being retained in the paint film, especially when the waxed film has been exposed to sunshine. A combination of wax and hot sunshine can produce a condition like blooming or dulling.

1. Do not attempt to use a wax polish but apply a spirit/solvent wipe; several applications may be needed.
2. Lightly compound the area.
3. Repeat solvent wipe and then use a liquid polish with a clean cloth.
4. Continue with solvent wipe, light compounding and polishing at each reappearance and as long as the wax continues to bleed from the surface.

Water leak diagnosis

The repair of body shell damage invariably requires the sealing of the engine, passenger, load and luggage compartments. In some instances the vehicle may be returned after body repair with complaints of water ingress. In these instances, the conditions under which leakage occurs should be established. These conditions will normally be:

1. When the vehicle is stationary in heavy rain.
2. When the vehicle is being washed with a hose.
3. When the vehicle is in an automatic car wash.
4. When the vehicle is mobile.

Further assistance in diagnosing leaks can be gained if:

1. Water can be seen to enter the car.
2. The water can be identified as clean or dirty.

Answers to these points may indicate whether the leakage is from body seams exposed to water thrown up by the road wheels or whether the upper body seams and/or weather strips are defective. To simulate the various driving conditions, it is necessary to use a hose pipe equipped with a means of varying the spray pattern and pressure. Such equipment can then be adjusted to give:

1. A high-pressure jet or spray of water simulating water thrown up by the road wheels on to the underbody and wheel arches.
2. A medium-pressure jet or spray of water simulating driving rain on the upper body panels and components.
3. A low-pressure jet or spray of water to test areas normally protected from direct contact with driving rain, such as underbonnet seams and grommets.

When using a hose at medium or high pressure, the nozzle must be held at least 1 m from the area under test. When diagnosing leakage from the aeroflow ventilation or the heater box, water must not be sprayed directly into the component under test.

As an alternative to diagnosing by water hose, ultrasonic equipment is available. This equipment utilizes a battery-operated high-frequency sound transmitter and receiver, the signal being received either through headphones or by warning lights. Instructions and test procedures for using this type of equipment are supplied with the equipment. Whether ultrasonic equipment or a water test hose is used, diagnosis must follow a logical sequence:

1. Remove the interior trim materials from the suspect area.
2. Remove all traces of water from the area under investigation.

3 Starting at the lowest point, spray water at medium pressure on the upper body exterior seams and seals and at high pressure on to the underbody seams and wheel arches. This operation should be carried out preferably by two people, one operating the hose from outside the vehicle and one inside the vehicle checking the interior for water entry with the aid of a suitable light.
4 Move the spray, or jet, of water slowly upwards until a point of water ingress is located. Repair that point (a temporary repair may sometimes be advisable at this stage, e.g. the application of sealer where braze or weld is required as a permanent repair).
5 Continue the diagnosis, repairing any source of water entry until the highest point of the suspect area is reached and all possible leak locations have been checked.
6 Complete the repairs where temporary seals have been made and thoroughly retest the area.
7 Replace the trim materials previously removed to gain access to the suspect area.

Coachlines and self-adhesive decor

Preparation
1 Wash complete area and surrounding areas with a solution of water and water-soluble cleaner or mild detergent to remove dirt and traffic film.
2 Remove self-adhesive lines and decor. These plastic materials are temperature sensitive, becoming brittle and extremely difficult to remove when cold.
 (a) It is necessary to warm the decor or coachline to soften the adhesive and render the material more flexible. A hot-air blower may be used to apply gentle heat to the line as it is peeled off.
 (b) When removing the decor ensure direction of pull is at 90 degrees to the surface; alternatively, roll the decor off. If the material splits into small pieces, alter the direction of pull.
 (c) Remove any adhesive remaining on the paint surface using an approved adhesive cleaner. A liquid compound may also prove successful.
 (d) If the panel is to be refinished, ensure that adhesive is removed during the flatting operation.
3 Thoroughly degrease the area with a solvent wipe. Ensure that new paint finishes have hardened before applying decor.
4 Ensure the surface is at a temperature of not less than 16 °C.
5 Denib the area and ensure all dirt is properly removed using a very fine burnishing pad or 1200 grit abrasive used dry.
6 Lightly compound area to which the decor is to be applied.
7 Spirit wipe to remove traces of compound using a clean cloth.
8 Thoroughly degrease and spirit wipe inside panel edges for wrapround of decor – this is vital for proper adhesion.
9 Check the decor to ensure that it is correct for the application and of sufficient length to allow at least 3 mm (0.25 in) overlap for wrap-round at edges of panels.

Application

Narrow coachlines Mark start and end position using a strip of masking tape. Tack off panel.

1 Peel back 75 mm (3 in) of the backing paper. Do not handle the adhesive after peeling off the backing strip, especially at the ends and tapered sections etc.
2 Tear off 25 mm (1 in) of the backing paper from the peeled end, then attach this across the exposed adhesive at the start of the tape to prevent soiling the end when positioning.
3 Align the coachline to the start mark on the panel indicated on the piece of masking tape previously applied and to the correct alignment of the panel, and then press exposed adhesive start end firmly onto panel.
4 Peel back a length of the backing paper to allow the coachline to be gently pulled straight into correct alignment without hand-

ling the adhesive and then touch down on to panel. Do not press down until complete line has been applied and rechecked for alignment. The length of backing to be removed at any time must not greatly exceed the length of coachline that can be accurately aligned and this will depend on the contour and curves of the panels and skill of the operator. As a general rule straight lines will be applied in a long length before touching onto panel. Curved lines will need short lengths and frequent touching down.

5   Do not stretch the coachline while applying.
6   If it is necessary to further remove the line use short pieces from the discarded backing paper to protect the adhesive at the points of handling.
7   Remove any pieces of backing paper used as adhesive protection. Check for freedom from kinks. After checking the alignment, work from the centre of the line out to the edges, press the line firmly along its length using a rubber roller, soft plastic applicator, or a suitable pad.
8   Wrap ends around panel edges and press down firmly.
9   Remove the protective paper if applicable. Press the line firmly along its length using a clean cloth folded to form a pad.

When the decor is a complex shape Offer the decor up to the surface and apply strips of masking tape to mark its end positions and hold in position on the panel with strips across it. Cut these strips to remove the decor leaving the pieces stuck to the panel and the decor for alignment during application. Tack off surface.

1   Remove 75 mm (3 in) of backing paper at thickest point and tear this off. Replace a part of this piece to protect the exposed adhesive edge from soiling while handling.
2   Align the decor to the masking tape pieces on the panel previously applied and to the longest straight edge of the panel and hold in position using strips of masking tape across it.
3   Touch down exposed adhesive on to panel. This may need repositioning. Press lightly.
4   Pull decor straight and apply light tension without stretching the decor. Use the edge of a hard plastic applicator as it is touched down on to the panel to ensure that it is kept flat. Cut the strip of masking tape used to hold the decor in place. Peel back a length of the backing paper to allow the decor to be gently pulled straight without handling the adhesive and then touch down on to panel. Do not press down. Ensure that the decor is not allowed to twist or stretch at its edges, as this will result in creases.
5   Remove any pieces of backing paper used as adhesive protection. Check for freedom from kinks. After checking alignment, work from the centre of the decor out to the edges, pressing down firmly along its length using a rubber roller, soft plastic applicator or suitable pad.
6   Wrap ends around panel edges and press down firmly.
7   Remove the protective paper if applicable. Press the line firmly along its length using a clean cloth folded to form a pad.

When the decor is long Long decor is applied as for coachlines but needs a line of masking tape as a guide edge.

Air blisters may be removed by puncturing with a very sharp pin and then working the air out from the edges into the pinhole.

9 Refinishing systems

Identification of original finish

It is essential that the manufacturer's original paint type is correctly identified in order to ensure compatibility of the refinish system. Unknown finishes can be identified by using cellulose solvent and/or scraping with a sharp blade. Ensure that tests are confined to the area to be refinished.

Cellulose solvent method

Apply solvent by rubbing onto a small area with a clean cloth folded into a small pad. Repeat the application to allow for a short reaction time before the solvent evaporates. Avoid excessive application of solvent as this may affect underlying coats of paint. Cellulose solvent will soften and dissolve the following two paint types:

1. Nitro-cellulose (air-dried lacquers).
2. Thermoplastic acrylics.

The solubility of the paint will be indicated by colour transfer on to the cloth.

Note: Oleoresins and some polyurethane synthetics will wrinkle when exposed to heavy applications of cellulose solvent. If the paint is not dissolved by a cellulose thinner, it will be of a thermosetting acrylic, melamine alkyd or a synthetic type. These may be identified by scraping the film.

Scraping with a sharp blade

Thermosetting acrylics will scrape off as a fine powder. Melamine alkyds will take the form of threadlike chips, and synthetics generally scrape off in the form of smooth chips. Clear-coat finishes are easily identified when rubbed with a piece of fine, dry abrasive paper as the resulting scratches will appear as white lines and a white dust will form on the abrasive.

Cellulose colour system: nitrocellulose synthetic, nitrocellulose, cellulose lacquer

Cellulose paints are a combination of cellulose and synthetic resins, with various pigments. The resins are also called varnish, binder, or the paint vehicle, and contribute to the adhesive qualities of the paint. The pigments are essentially the solids content of the paint. During drying, the pigment and the resins become the sole ingredients of the final paint film. These two components are made workable for spray application by the addition of a strong solvent or thinner.

The solvent is a formulation of various hydrocarbons of different density and hence evaporation rates. Each particular solvent blend, with specific rate and sequence of hydrocarbon evaporation, will influence the drying and consequently the final properties and appearance of the 'cured' film.

Celluloses are not recommended for refinishing thermosetting plastic acrylic (TPA) because of inferior colour retention and tendency to fade or chalk quicker than TPA. There is also the possibility of lifting at feathered edges or of incompatability defects if used directly on TPA. Attempts to insulate the TPA from the effects of the solvent in cellulose paints, by using a primer or a sealer coat, will not necessarily guarantee a satisfactory refinish. Cellulose finishes are not suitable for use over oleoresin, synthetic and some polyurethane paints. Where the suitability of cellulose is uncertain for refinishing, test a small area requiring refinish by saturating with cellulose thinner. If the surface reacts with the thinner as indicated by any undesirable effect, it will be unwise to refinish using cellulose products.

The combination of the strong solvents and low solids content render celluloses prone to sinkage and low scratch-filling ability. High paint usage becomes evident in order to obtain a satisfactory final paint finish unless attention is given to careful, thorough and fine abrasive preparation. Failure to 'feather' original paint edges or prepare the substrate properly prior to application of cellulose colour will result in outlines in the finish and the possibility of lifting. Additional care is necessary when initially applying over feathered edges of compatible synthetic paint or synthetic-type primers.

When using cellulose products, high gloss levels may be achieved by additional quantities of colour, flatting and polishing. The excellent response of cellulose colour finishes to polishing enables rapid and easy rectification of minor defects. Cellulose paints afford exceptional advantages in ease of use, application and economy with high 'turnover' when used for local resprays, small areas and single-panel rectification, etc.

Thinning

The final appearance of the colour film 'from the gun' is greatly dependent on the use of the correct high-quality thinner. Although the 'spraying' viscosity will depend on the operator, equipment and air temperature, a paint to thinner ratio of 40:60 by volume, i.e. 2 parts paint to 3 parts thinner, provides a viscosity of approximately 23 seconds (BSB4 flow cup at 21 °C).

Application

Spray a test area to check atomization and spray pattern. A spraying pressure at the gun of 3.85–4.55 bar (55–65 psi) is recommended. The first colour coat must be applied thin and wet. Allow 5 minutes flash-off and follow with a full single coat. Allow 10 minutes minimum flash-off (longer drying between coats may reduce sinkage). Apply either one or two further single coats or a double coat depending on the covering power of the colour (opacity) and the spraying viscosity. Single coats will allow rapid drying while a double coat will promote good flow with freedom from dry edges. When refinishing large areas in cellulose it is recommended that the final colour coat is applied as a double coat.

Metallic colours

Always spray a test panel in order to establish the pressure, viscosity and technique which will provide the closest match. (The final appearance of the colour can be altered by using different spraying techniques and it is not always necessary to use a tint to obtain a match.) Some cellulose-based metallics will require additional colour coats to obtain the correct opacity (coverage). When spot respraying with metallics, the use of clear blending additives are recommended. A longer flash-off may be necessary with metallics.

Drying

The film will air dry to a hard condition in 3–6 hours depending on the thinner used, film thickness and air temperature. Complete hardening requires approximately 16 hours air drying. After a 15–20 minute flash-off cellulose can be force dried in 30 minutes at temperatures up to 70 °C. After cooling it will be fully hardened.

Converting cellulose for stoving

Some paint manufacturers offer cellulose materials with the option of conversion to a low-bake system. This is achieved by using either a special thinner, instead of the usual cellulose thinner, or an additive.

Local respray

Approximate application viscosity is obtained with a paint to thinner ratio of 40:60 by volume. Apply a wet thin single coat and flash off for 5 minutes. Apply a further two single coats with a 10 minute flash-off between each coat overlapping the dry edges of the previous coat, and removing dry spray from the area before each successive coat. Reduce spraying pressure to 2.10–2.80 bar (30–40 psi). Thin further 50:50 by volume, spray a light coat to cover previous edges and apply mist coats to fade into surrounding finish. Special thinner is available for fade-out techniques and can be used for the final thinning instead of the normal finish thinner. Blending additives may also be used to promote fade-out.

Surface preparation

Economy of both time and material will be obtained by wet flatting the substrate or primer surface with no coarser than a 600 grade abrasive. Use 800 grade for preparation prior to a local respray. Attention must be given to correct feathering of paint edges. When wet flatted ensure that the primer has sufficient time to dry properly.

Careful surface preparation is vital when using cellulose to achieve quickest refinishing without using excessive material or polishing effort. Dry flatting with no coarser than 320 grade with machine or 400 grade stearate paper by hand may also be employed but will not produce as good an appearance as wet flatting.

Primers

There are two types of primer which may be used with cellulose: cellulose based and synthetic modified.

Cellulose primer–surfacer–filler

Current cellulose primer materials are dual purpose. They will have filling and surfacer characteristics as well as functioning as a primer. By varying the viscosity and method of application emphasis can be placed on either the surfacing or the filling capabilities. The potential rapid drying of cellulose products (in as short a period as 30 minutes) will enable high filling rates, but coats must not be applied too dry or too thick. Dry application and excessively thick coats result in porosity of the primer and, consequently, sinkage of the colour coats. Additional flatting will be required if primers are not applied properly.

Spray at a pressure of 3.50–4.55 bar (50–65 psi) at the gun consistent with correct spray pattern. It is essential to avoid application in cold atmospheres or conditions which may cause blushing (moisture condensation in the film). Blushing is difficult to detect in the matt finish of primers, and may cause poor adhesion or blistering of subsequent coats.

Apply cellulose primer in wet coats whether filling or surfacing. Quick drying for primer–surfacing will be obtained by applying single wet coats at a viscosity of 20–24 seconds BSB4 flow cup with 5–10 minutes flash-off between coats.

Primer–filling is obtained by applying at a viscosity of 25–32 seconds BSB4 cup. It is essential to first spray a thin, wet single coat and allow to flash off for 5–10 minutes prior to application of successive filler coats. High build rates are possible with these primers. It is essential that correct flash-off and proper through drying times are observed. Failure to allow correct through drying will result in sinkage of the cellulose colour coat. It must be noted that cellulose primer is permeable to water; for this reason, ensure that it is thoroughly dried out after wet flatting. Moisture remaining in the primer will cause blistering.

Synthetic resin primers

Note: This section does not include oleoresin-based primers.

Synthetic primers are equally suitable for use with cellulose finishes but are slower drying than the cellulose-based primers. Hardening requires from 1.5 to 10 hours at 20 °C before flatting is possible. Synthetics have excellent hold-out and high filling capabilities and present greater resistance to moisture than cellulose primers. The recommended spraying pressure at the gun is 3.15–4.20 bar (45–60 psi). Thin to a viscosity of 25–30 seconds BSB4 cup at 21 °C. Some variations of synthetic primers may only be suitable for single-coat application. These must be applied as directed by the manufacturer. The flash-off period between coats for synthetics will be longer than for cellulose-based products, ranging from 10 to 35 minutes. After final flash-off (15–35 minutes) force drying may be accomplished at temperatures up to 80 °C depending on the product. Dry flat with 320 grade or wet flat with 600 grade paper.

Stoppers

Two types of stoppers may be utilized – polyester for bare metal only, or cellulose for shallow filling between primer coats. Do not use polyester stopper between coats of cellulose materials. Always ensure that stoppers are flatted with the recommended grade of abrasive and are correctly primed prior to application of cellulose colour coats.

Etch primers

Do not use etch primers over polyester-type stoppers unless recommended by the manufacturer. Use of etch primers with the cellulose system increases its durability and with metal pretreatment provides the equivalent of manufacturer's electrocoat primer where this has been removed. Some paint manufacturers provide etch primer–fillers. These are dual-function etch primers and fulfil the role of primer-filler. A surfacer may be required over these prior to cellulose colour.

Acrylic colour system: air-dry acrylic, thermoplastic acrylic

Acrylic paints are formulated from a varnish or paint 'vehicle' of acrylic resins (a type of plastic resin) with various pigments (the solids content). The resin and pigment are made workable by the addition of a blend of hydrocarbons (solvent or thinner). The evaporation characteristics of the solvent determine the drying rate of the film, the surface finish and ultimately the gloss level. Polishing requirement and resulting operation times will be at a minimum when acrylic is sprayed in a temperature-controlled booth to reduce the effects of humidity and atmospheric conditions – maintain air temperature between 20 and 25 °C. Some polishing is often necessary to achieve an acceptable gloss level but this is readily achieved from a correctly sprayed acrylic finish. Also, exceptionally high gloss levels can easily be obtained from an acrylic lacquer.

Acrylic finishes must not be used over existing cellulose colour finishes unless recommended by the paint manufacturer (acrylic solvent has a softening effect on the cellulose with the possibility of crazing as the acrylic contracts during drying or later in service). Acrylic lacquers (thermoplastic acrylics or TPA) are distinguished from other 'acrylic' formulations by the property of curing solely by evaporation of the solvent.

Thinning

When acrylic lacquer is correctly thinned at a recommended paint to thinner ratio of 40:60 (1:1½) it will be at a very low viscosity of approximately 16–18 seconds BSB4 cup at 22 °C. Various thinners are available for different drying rates, set-up and flash-off. In conditions which create fast set-up, it is important that an appropriate thinner which promotes improved paint flow is used. Failure to thin acrylic sufficiently will result in 'threads' of solidified paint issuing from the fluid nozzle in the form of a 'cobweb'.

Application

A spraying pressure of 2.80–3.85 bar (40–55 psi) at the gun is recommended. Spray a test area first, particularly when the use of acrylic is unfamiliar. Application of acrylic is different from other types of refinish paints. When a coat has flashed off and set-up, smooth flow-out will not be induced by application of successive wet coats. It is essential that each coat is applied wet and smooth to obviate excessive polishing and to obtain a smooth final finish. Acrylics cannot be successfully applied if the temperature differences between surface, surroundings and/or paint exceed 5 °C. The spraying sequence must be planned to avoid dry edges when refinishing large areas. Do not apply colour coats in quick succession without adequate flash-off. Apply three to five single wet coats, allowing 5–10 minutes flash-off between each. Alternatively apply a single coat followed by one or two double coats with a 5–10 minute flash-off between coats. The number of colour coats required will depend on the colour and opacity and whether the finish is applied over primer or original colour.

Metallic colours

Metallic films can be built up in the same manner as solid colours, using single coats and for final application either a single or a double coat, depending on the metallic effect required to match the original finish. Clear acrylic may be used in conjunction with metallic colours to provide additional film thickness for final polishing when very high gloss is required.

Drying

Drying of acrylic coat at 21 °C will require 30 minutes to touch dry and 3 hours before polishing. Complete through hardening at 21 °C requires a 16 hour period. Alternatively, after final flash-off at 20–30 minutes, force drying may be used to accelerate through hardening. At a temperature of 60 °C the time needed for hardening can be reduced to 30 minutes.

Converting for low bake/force dry

Some air-dry acrylics may be converted for low-bake use by using a special converter thinner instead of the usual air-dry version.

Local respray

Apply three or four thin single coats, ensuring adequate drying between each coat. Thin by a further 50 per cent and apply two single coats, overlapping each previous coat. After thinning a further 50 per cent apply one or two thin, wet coats to the entire area using the fade-out method. Clear additives may be available for fading out colour. Retarder thinners may also be used to advantage for spot respray techniques. Polishing of an incompletely dried film will achieve a gloss but this will be lost by subsequent solvent evaporation.

Surface preparation

Surface preparation includes use of the recommended primer, adhesion promoting products (isopropyl alcohol wipe when specified by the manufacturer) and operator techniques, besides careful and fine abrasive flatting of the surface.

When acrylic colour is to be applied directly over an original or existing thermoplastic acrylic (TPA) colour coat it should be wet flatted with 800 grade paper. The primer should be either dry flatted with 320 grade or preferably wet flatted with 600 grade. Wet flat with 800–1000 grade for local respray.

Attention must be given to correct feathering of paint edges and careful surface preparation with regard to paint edges is vital to eliminate risk of film defects.

Primers

If primer film thickness is excessively reduced by flatting, the colour finish may be impaired. Primers may be either a recommended cellulose type or a synthetic/acrylic type. Ensure that primer is properly hardened before flatting. Where an original acrylic (TPA) finish has broken or is sanded to bare metal it is imperative that the primer is compatible to avoid lifting.

Cellulose primer–surfacer–filler

Cellulose primer materials for use over acrylics may be used as fillers and surfacers depending on their characteristics. Spray with a pressure at the gun of 3.50–4.55 bar (50–65 psi) consistent with correct spray pattern. Whether filling or surfacing always apply the initial coat thin and wet. Avoid application in cold atmospheres and allow to dry thoroughly after final flash-off. Rapid drying of primer-surfacing will be obtained by applying single wet coats at a viscosity of 20–23 seconds BSB4 cup with 5–10 minutes flash-off between coats. Primer–filling is achieved by application at a viscosity of 25–32 seconds.

Synthetic resin primers

Note: This section does not include oleoresin-based types.

Synthetic resin primers are equally suitable for use with acrylic finishes but require from 1.5 to 10 hours at 20 °C before flatting. Synthetics have excellent hold-out and high filling capabilities. They also present greater resistance to moisture than cellulose primers of equivalent film thickness. The recommended spraying pressure at the gun is 3.15–4.20 bar (45–60 psi). Thin to a viscosity of 25–30 seconds BSB4 cup at 21 °C. The flash-off period with these products is longer than for cellulose-based primers. Apply as directed by manufacturer as some variations are intended for application in single coats only.

After appropriate final flash-off (15–35 minutes) force drying may be accomplished at temperatures up to 80 °C depending on the product. Dry flat with 320 grade or wet flat with 600–800 grade abrasive paper.

Stoppers

Polyester stoppers

Use polyester stoppers for bare metal application or as recommended by the manufacturer for use with acrylic. Do not use polyester stoppers between coats of acrylic materials or over original thermoplastic acrylic (TPA).

Cellulose stoppers

Do not apply cellulose stoppers directly to TPA. These stoppers are only suitable for use with acrylic when applied between coats of primer or on to bare metal. Cellulose stoppers must be sealed by a primer film prior to acrylic colour application. Always ensure that stopper is flatted and finished with fine grade abrasive used wet. Apply three coats of primer prior to application of acrylic colour coats.

Etch primers

Do not use etch primers over polyester-type stoppers when these have been applied to the bare metal. Use of etch primers with the acrylic system increases its durability and, with the utilization of metal pretreatment chemicals, provides the equivalent of the manufacturer's electrocoat where this has been removed. It is essential that etch primers are not applied over the original TPA paint edges or finish. Give particular attention to local resprays where bare metal requires etch treatment and ensure that contact with the feathered edges of the original TPA is avoided.

Acrylic modified synthetic (2K, two-pack) colour system: isocyanate hardened paints, 2K acrylics, two-pack acrylic enamel

These paints are a blend of acrylic and synthetic resins (alkyds) and are similar to the thermosetting acrylic (TSA) production synthetics. Unlike production TSA, the properties of an isocyanate catalyst are utilized for hardening rather than the high-bake processes employed by vehicle manufacturers. The mild solvents used with these paints permit their use over most substrates and suitability for refinishing most production finishes. Although polishing is not normally required, acrylic modified paints respond extremely well to removal of film defects and surface imperfections by flatting, compounding and polishing. It is essential to use these paints only in a spray booth. A temperature-controlled booth will provide maximum benefit and efficiency of the system and application at temperatures between 20 and 25 °C is generally recommended by the paint manufacturers. When refinishing over an existing paint film the vehicle manufacturer's recommended limit must not be exceeded. If there is a possibility of exceeding the maximum thickness, the film must be sanded to allow for the additional refinish material.

Thinning

Catalysed paints may be sensitive to the sequence in which the components are mixed. Use only recommended thinning solvents, component ratios and procedure. The finish achieved with isocyanate catalysed paints is greatly influenced by application viscosity; 20–24 seconds BSB4 cup at 21 °C is generally acceptable but minor adjustment to the viscosity may be made to suit the operator using the appropriate thinner. It is essential to recognize the limitation imposed by the pot life of the mixed colour finish with regard to the influence of atmospheric conditions and air temperatures. The pot life is the safe workable time after the catalyst is added to the paint.

Application

Isocyanate paints must be used only in a suitable spray booth with proper air circulation and operator respiratory equipment.

Static charges may be induced by any rubbing motion of preparatory operations; as a result it may be necessary to earth the surface to reduce dirt attraction.

Do not mix the components until the paint is required for use. Set the air pressure at the gun to 3.85–4.90 bar (55–70 psi) (this will depend on the operator, particularly with these refinish materials). The recommended temperature for paint, surfaces and atmosphere is 20–22 °C.

Although the overall hardening of acrylic modified synthetics is not excessively affected by atmospheric variations, temperature variations have a particularly noticeable effect on the viscosity, flash-off and set-up times during spraying. This must be accommodated by altering the rate at which the product is applied to the surface or by using a thinner and hardener to suit.

Spray all hard edges first. Do not apply thick colour coats. First apply a thin light coat and allow this to flash off for 2 to 3 minutes. Follow with two single wet coats, allowing 5 minutes flash-off between each. Alternatively, apply three single coats or a single coat followed by one double coat with a 5–10 minute flash-off between coats. The number of colour coats required will depend on the colour, opacity and viscosity of the paint.

Metallic colours

Synthetic metallic colours require application at a lower viscosity (thinner) than solid colours. The lower viscosity promotes quicker set-up and prevents excessive separation of the aluminium flakes which would result if the paint was applied at high viscosity. Additional coats may be necessary to give full opacity. A viscosity of 16–19 seconds BSB4 cup at 21 °C will achieve a normal metallic tone and alterations can be made to vary this as required.

Metallic films are built up in the same way as for solid colours, that is in two single coats, then applying either a single or a double coat, finishing with a dusting technique, depending on the metallic effect required to match the original.

Drying

Normal drying at 21 °C will require 15 minutes to touch dry and approximately 1 hour before handling. A minimum period of 6 hours is necessary before the paint film can be subjected to service conditions. Full curing of the paint film takes about 4 days. Alternatively, after a minimum final flash-off of 15–20 minutes, force drying may be used for more rapid through hardening. At a temperature of 60 °C complete hardening time can be reduced to 30 minutes.

Converting for low bake/force dry

Some manufacturers of the acrylic modified synthetic paint system offer the option of conversion for use specifically as a low-bake system. This adaptation is effected by using specially formulated additives and thinner.

Local respray

Acrylic modified synthetics require careful and skilful applications to achieve satisfactory fade-out of paint edges. Thin the colour to 18–20 seconds BSB4 cup at 21 °C. A pressure of 2.80 bar (40 psi) at the gun is recommended for the first coats to obtain colour covering. Set the spray pattern to a narrow fan and reduce the paint flow until the paint is consistently distributed and atomized correctly. Apply a thin mist coat and flash off for 3 minutes. Apply two single coats, ensuring adequate drying between each coat. Particular attention must be given to the removal of dry edges between each successive coat. Reduce the spraying pressure to approximately 2.10 bar (30 psi). Thin the paint by a further 50 per cent and apply two single coats, overlapping each previous coat and allowing each to dry before the successive coat. Remove dry edges with 1000 grade paper. After thinning by a further 50 per cent apply a mist coat to the entire area using fade-out method to wet and dissolve dry spray at edges. Before polishing, ensure that the film is dry and properly cured.

Surface preparation

Fine flatting prior to colour coats is necessary by wet flatting with 600 grade paper or dry flat with 320 grade and an orbital machine sander. The thickness of an existing paint film must be reduced at this stage to allow for the additional refinish colour film. If the surface is prepared with coarse abrasive, additional colour material and an excessively thick film will have to be applied to achieve a satisfactory appearance. Thick films are not conducive to film durability. For local respray wet flat with 800 grade and thoroughly feather all paint edges to avoid outlines in the finish.

Primers

The primer can be either a cellulose alkyd or a synthetic alkyd but must be chosen and used as recommended by the paint manufacturer. Failure to use correct primers will result in flaking or blistering of colour coats.

Cellulose primer–surfacer–filler

It must be noted that some cellulose products are not recommended for use with synthetic finishes. Most primers can be used as primer–filler or primer–surfacer according to the method of application. High build is possible in a short time as cellulose primers dry rapidly.

Spray with a pressure at the gun of 3.50–4.55 bar (50–65 psi) consistent with correct spray pattern. Each coat must be applied wet to avoid the possibility of porosity of the primer with subsequent sinkage of the colour coats. Cold atmosphere or conditions which may give rise to moisture condensation in the film must be avoided. Ensure after wet flatting primer that it is thoroughly dried out. To obtain quick drying for primer-surfacing apply single wet coats at a viscosity of 20–23 seconds BSB4 flow cup with 5–10 minutes flash-off between coats. Primer–filling is obtained by applying at a viscosity of 25–32 seconds BSB4 cup. It is essential to first spray a single thin wet coat and allow to flash off for 5–10 minutes prior to application of successive filler coats.

Synthetic/alkyd primers

Note: Some manufacturers of acrylic modified synthetic paint supply synthetic primers which they recommend as the only basis for the colour coat.

Synthetic primers will require from 1.5 to 10 hours at 20 °C before flatting. Synthetics have excellent filling capabilities and moisture resistance. Use a spraying pressure at the gun of between 3.15 and 4.20 bar (45–60 psi). Thin to a viscosity of 25–30 seconds BSB4 cup at 21 °C, this will give the recommended dry film thickness per single coat. Apply as directed by the manufacturer as some varieties may be suited to application in single coats only. The flash-off period with these products must be observed and will often be longer than the requirement of cellulose primers. After the appropriate flash-off (15–35 minutes), force drying may be accomplished at temperatures up to 80 °C (the temperature and times will depend on the product). Dry flat with 320 grade or wet flat with 600 grade abrasive.

Stoppers

Cellulose and polyester-type stoppers may be used with synthetic paints. Use polyester stopper only on bare metal unless recommended for use between coats of primer. Always ensure that stopper is correctly applied and hard before it is flatted with a fine grade abrasive paper. Use an abrasive of the same grit size as used for flatting the primer. Stoppers must be primed before colour application.

Etch primers

Do not use etch primers over polyester-type stoppers. Use of etch primers and metal pretreatment provides the basic equivalent to the manufacturer's electrocoat primer where this has been removed.

Synthetic colour system: synthetics, polyurethane synthetic, alkyd synthetic, oleoresin synthetic, low-bake system

There is a wide range of synthetic paints, some of which are modified with a special hardener or thinner to speed drying, curing and the general refinishing process. The solvent used with these refinish materials will not normally affect a substrate; for this reason synthetics may be used over most other finishes. The excellent scratch-filling and covering abilities of synthetic paints are attributable to the high cohesive properties of the alkyd resin and high pigment (solids) content. A solvent is added to the combination of resin and pigment to make it workable for manufacture and spray application.

Initial set-up occurs as a result of the evaporation of the solvent and this is affected by temperature and climatic variations. The paint film hardens only by an oxidation process which remains effective despite temperature and atmospheric fluctuations (although the time required for hardening may differ as a result of such variations). Set-up, drying and hardening times can be reduced by application at temperatures between 21 and 25 °C and, with the paint manufacturer's assent, stoving may be employed.

A temperature-controlled spray booth is recommended not only to promote efficient refinishing but also for prevention of dirt in the finish and contamination of surrounding surfaces by wet spray dust during spraying operations.

Thinning

It is necessary to check the manufacturer's instructions as the properties of synthetic products vary widely but an average viscosity of 25–30 seconds BSB4 cup at 21 °C is generally acceptable. Some synthetics may be hot sprayed by heating the paint to approximately 60 °C (this effectively reduces the paint to the correct application viscosity). The temperature of the paint must be maintained during application when spraying in the normal manner.

Application

Set the air pressure at the gun to 3.85–4.55 bar (55–65 psi). It is important to recognize that relevant compensation for viscosity changes due to temperature variation will be required, using an alternative thinner and possibly a different method of application. The recommended temperature for paint, surfaces and atmosphere is 20–22 °C unless hot spraying. Spray all hard edges first with a light coat and flash off. Apply one or two very thin, light coats allowing to flash off for 15–30 minutes between coats. Apply a final single wet coat. If hot spraying, a single wet coat will normally provide the required finish.

Metallic colours

Metallic films are applied by spraying two or three light coats. First apply a mist coat. Apply the final coat lightly but wet to enable further adjustment to the particle disposition by further dusting technique as required. Additional coats may be necessary to give full opacity or to match the metallic effect in the original film.

Drying

At 21 °C normal air drying, the film will be dust free after approximately 20–60 minutes and up to 4 hours before it may be touch–dry or handled. A minimum of 12–18 hours may be necessary before it can be put into service. The film may require as long as 5 days (when air dried) before it can be exposed to normal service conditions. After final flash-off of 20 – 30 minutes, force drying at temperatures up to 60 °C may reduce hardening time to 30 minutes. Do not attempt to stove a synthetic paint unless approved by the manufacturer.

Low-bake synthetic

There are some synthetic paints which can be converted specifically to low bake use by using a low-bake converter thinner. The application and peculiarities will be determined by the product.

Local respray

It is difficult to achieve satisfactory fade-out of the paint edges after using synthetic refinish material for a local respray. Some manufacturers may offer a specific thinner or additive to facilitate blending of the paint edge. Thin the colour to 24–26 seconds BSB4 cup at 21 °C. Adjust the air pressure to approximately 2.80 bar (40 psi) at the gun. Reduce the paint flow and spray pattern to a narrow fan with consistent distribution of paint and correct atomization.

First apply a thin mist coat and flash off for 10 minutes. Follow with a single light coat and allow to set up for 20–30 minutes. Reduce the spraying pressure to approximately 2.10 bar (30 psi). Thin the paint to 22 seconds BSB4 cup and apply a single dust coat to cover previous dry edges. Allow to flash off. Add a further 25 per cent thinner and mist coat the entire area using an arcing motion of the spray gun until dry paint edges are wetted (use a fast thinner if available).

Surface preparation

Fine flatting prior to colour coats is necessary by wet flatting with 400–600 grade or dry flat with 280–320 grade paper by orbital machine sanding. After fine preparation in this way, a consistent thin colour film can be applied without any risk of flatting marks appearing in the colour finish. Thoroughly feather all paint edges to avoid outlines in the finish.

Primers/undercoats

Undercoat (ground coat or glaze coat) of a particular colour may be necessary to supplement the finish colour. These may be a mix of finish colour and a light synthetic primer, or obtained as a special formulation, as recommended by the paint manufacturer. It is essential that only recommended primers are used for the synthetic system. These will usually be of a synthetic type. Some paint manufacturers supply special non-sand synthetic primers and undercoats which they recommend as the only basis for the colour coat. Synthetic primers are comparatively slow drying, requiring from 1.5 to 10 hours at 20 °C before flatting.

The recommended spraying pressure at the gun is from 3.15 to 4.20 bar (45–60 psi). Thin to a viscosity of 25–30 seconds BSB4 cup at 21 °C and apply one to three coats as directed by the paint manufacturer. Many synthetic primers can be applied as a double coat. The flash-off period with these products must be observed. Dry flat with 280 grade paper or wet flat with 400–600 grade unless the primer is intended for use as a non-sand or wet-on-wet basis (when colour is applied immediately after flash-off).

It may be possible to force dry a synthetic primer at temperatures up to 80 °C but only after appropriate flash-off (15–35 minutes) depending on the product.

Stoppers

Polyester stoppers are not prone to sinkage and, as a result, are more commonly used than cellulose types with synthetic paints. Use polyester stopper only on bare metal unless it is designed for use between coats of primer. Always ensure that stopper is correctly applied and hard when flatted. Use a fine grade abrasive paper of the same grade as that used for flatting the primer. If cellulose stopper is applied it must be primed before colour application.

Etch primers

Do not use etch primers over polyester-type stoppers when these have been applied to the bare metal. Special etch primers are available for use with synthetics to improve durability. Etch primers must not be applied over synthetic paints. The use of etch primer and metal pretreatment is essential to provide the equivalent of the vehicle manufacturer's electrocoat primer when this has been removed.

Clear coat metallics

The clear coat metallic paint technology offers significant advantages over conventional single-coat metallics. These advantages apply both to original production and refinish systems and can be summarized as follows:

1 Increased durability.
2 Improved metallic appearance with availability of clearer, more colourful pigments with greater cosmetic appeal for light-coloured metallic finishes and marked contrast between top and side tones ('flip').
3 Image clarity, which is basically a means of determining the quality of the surface finish, is better defined.
4 Aesthetically pleasing, high gloss.
5 Easier to apply and colour match.

In order to obtain these advantages, particularly for the refinisher, there are several points which must be considered:

1 Attention must be given to correct substrate preparation, sanding, degreasing, primer application and fine flatting of primer, use of tack cloth and general freedom from dust, etc. as application of clear coat metallics may exaggerate improper processes, poor technique and contamination.
2 There will be an additional time requirement for fine wet flatting with 800 grade abrasive and for the application of base coat colour and clear coat, particularly regarding the flash-off times which must be observed.
3 Depending on the choice of material and refinish system, polishing may be necessary to achieve the required gloss level, especially when a TPA clear coat is used.
4 There may also be a relatively higher product consumption depending upon the system used, but in general the TPA product requirement will be greater than that of two-pack/2K acrylic types.
5 Base coat should be a thin layer, consistent with full opacity; avoid excessively thick application.

There are several systems available for refinishing clear coat metallic finishes:

1 *Base coat* cellulose.
   Clear coat thermoplastic acrylic (TPA).
2 *Base coat* cellulose.
   Clear coat two-pack/2K (isocyanate hardened) acrylic modified synthetic.
3 *Base coat* thermoplastic acrylic (TPA).
   Clear coat thermoplastic acrylic (TPA).
4 *Base coat* thermoplastic acrylic (TPA).
   Clear coat two-pack/2K (isocyanate hardened) acrylic modified synthetic.

5 *Base coat* acrylic modified synthetic (without isocyanate hardener).
  Clear coat two-pack/2K (isocyanate hardened) acrylic modified synthetic.
6 *Base coat* two-pack/2K (isocyanate hardened) acrylic modified synthetic.
  Clear coat two-pack/2K (isocyanate hardened) acrylic modified synthetic.
7 *Base coat* synthetic acrylic (PUR-polyurethane), polyester, etc.
  Clear coat thermoplastic acrylic (TPA) or two-pack/2K.

Always consult the product manufacturer's data and information.

10 Paint defects: cause, prevention and repair

This chapter gives paint failures and complaints which can generally be traced to improper paint handling and painting techniques encountered in the refinishing trade. The brief description of their cause, their prevention and their repairing may help the painter.

Adhesion

This section covers poor adhesion, peeling, flaking and blowing off.

Cause

1 Presence of any foreign material on surface prior to painting, such as wax, grease, silicones, oil, water, rust, solder flux and soap. (These foreign substances may come from many sources including waxes and polishes, compounds, detergents, compressed airline, hands, petrol, reducers and atmosphere.)
2 Use of wrong primer for the metal.
3 Improper use of, or no use of, metal pretreatment and rust remover on steel, aluminium or fibreglass.
4 Use of cheap thinner, insufficient thinner, or too high air pressure for primer.
5 Recoating of primer–surfacer without allowing sufficient time for primer to dry.
6 Insufficient flatting of surface before painting.
7 Application of coating to surface which is too hot or too cold.
8 Films too thick.

Prevention

1 Before starting metal work, wash area thoroughly with metal pretreatment and rust remover.
2 Neutralize solder flux with ammonia or baking soda solution, followed by weak vinegar solution and water rinse.
3 Wet flat or dry flat old finish *thoroughly*. Dry sand metal area.
4 Wipe entire area with a wax and grease remover.
5 Use metal pretreatment and rust remover on bare metal.
6 Dry surface with *clean* rags or chamois, and compressed air.
7 Prime bare metal areas within 30 minutes to prevent start of rusting.
8 Paint material, surface, and room should be at relatively equal temperatures, preferably between 16 and 24 °C.
9 Follow directions as to type and amount of thinner and air pressure.

Repair

The only method of repairing a paint job which shows signs of poor adhesion is to remove the old finish and repaint.

Bleeding

Cause

Soluble dyes or pigments in old finish, on surface, or in undercoat dissolve in solvent of new colour, and seep through the new finish colour.

Prevention

1. Before repainting a red or maroon colour with any light colour, check the old finish for bleeding by coating a small area with the light colour (or white). Bleeding will generally appear in a few minutes if the old colour is of a bleeding nature.
2. Do not allow spray dust from a bleeding colour to fall on other jobs.
3. Clean all equipment thoroughly after using a bleeding colour.

Repair

1. The best way to repair a bleeding area is to remove the paint in this area and repaint.
2. Mild cases of bleeding can sometimes be repaired by the use of lacquer sealer.
3. The customer may settle for a darker colour, or a red or maroon, which will not show the bleeding.

Blistering

This section includes pimpling, bubbling, pinholing, pitting and pock marks.

Cause

1. Usually caused by moisture becoming trapped between metal and undercoat, or between undercoat and colour coat, expanding, and forming small to large rounded blisters.
   Other causes or contributing factors are oil, dirt, rapid changes in temperature, large temperature differential between surface and paint being applied, cheap thinner, salts from water used for flatting.
2. The water causing the blistering may come from the flatting operation or condensation caused by the cooling effect of spraying. Water or oil may come through the compressed air line.
3. Since no paint film is completely impervious to water or water vapour, some moisture often gets under the film. When this occurs, heat (from sunshine, etc.) will vaporize the moisture, and, if the pressure developed is great enough, blisters will form. This type of blistering is more prevalent in rainy or humid conditions.
4. Degreasing with petrol often results in blisters forming in the finish.
5. Cheap, fast thinner; insufficient thinner; too high air pressure; and dry spraying of undercoats (causing porosity and air pockets in the undercoat film) can cause blistering.
6. Insufficient drying time and too heavy application of the undercoats may trap solvent, which later escapes to cause blistering of the colour coat.
7. Application of heat to a film containing volatile solvents.

Prevention

1. Check compressed air regularly for oil and water. One method of checking is to blow the air against clean glass or white paper. Inspect for oil or water droplets.
2. After wet flatting, blow out all cracks and crevices thoroughly with clean dry compressed air.
3. Do not allow water droplets to dry by themselves. Always wipe off visible water to avoid chemical deposits.
4. Allow sufficient time after wet flatting and blowing off for undercoat to dry out. On particularly humid days, several hours may be required.
5. Use a good thinner to avoid condensation of atmospheric moisture on surface.
6. Keep air pressure as low as possible consistent with good atomization as in information manuals.
7. Allow metal and paint to come to equal temperatures, 16–24 °C, before painting.
8. Apply undercoats in thin, wet films, allowing sufficient flash-off time between coats to avoid trapping solvents.

9 Do not apply heat to heavy films which contain volatile solvents.

Repair

1 If damage is extensive and severe, paint must be removed down to undercoat or metal, depending on depth of blisters. Repaint.
2 In less severe cases, blisters may be flatted out, and repainted.

Blushing (blooming)

Cause

A milky and sometimes dull appearance formed in paint films during or immediately after spraying, caused by precipitation of the components of the film, due to condensation of moisture on the wet film. Fast evaporating thinner, high air pressure, draughts, and rainy or humid weather all lead to this condition.

Prevention

1 Avoid spraying on rainy or extremely humid days.
2 Use high-grade thinner.
3 Use a retarder on humid days.

Repair

1 In extremely bad cases, respraying using a high-grade thinner.
2 In minor cases, spraying a coat of high-grade thinner or retarder will redissolve the blushed film and restore normal appearance.

Bronzing

A not too common condition, seen more often in some dark blues, reds, maroons and rarely in other colours, having the appearance of a fine coloured chalk or bronze coloured film on the surface. Easily wiped off, the condition appears again in a short time.

Cause

1 A peculiar combination of certain pigments and film-forming ingredients is responsible.
2 In a few cases, overheating of reds and maroons by hot spraying or wheel burning has been known to cause this condition.

Prevention

The painter has little or no control over this rather rare condition.

Repair

1 Light hand polishing with a mild liquid polish will remove the bronze. Frequent washing and light polishing will maintain good appearance.
2 Permanent elimination of the condition usually requires repainting.

Chalking

See also *gloss*.

Cause

Chalking is the result of normal destruction of a paint film through the effects of weathering, mainly due to ultraviolet light and moisture. All paint films will eventually chalk, giving the appearance of a dull, powdery, usually light-coloured surface.

It must be pointed out that the degree of chalking present on a finish is a function of many things, including the colour of the paint, age, care given (washing, polishing, whether kept in open or covered), severity of exposure (function of location and weather), as well as the actual paint used.

Complaints of excessive chalking can usually only be justified by comparing the finish in question with others of the same colour, age, care, and exposure.

Prevention

1. Use finishes of reliable manufacturers.
2. Do not use special additives to gain advantages such as quicker drying, harder films, better gloss, etc. except where specifically recommended by the paint manufacturer.
3. Follow thinning and application directions of the paint supplier, applying normal film thicknesses of undercoats and colour coats.
4. Avoid strong soaps and cleaning compounds for maintenance washing of car finish.

Repair

1. Light chalking requires only a light polishing to restore the original lustre.
2. Heavy chalking requires compounding and polishing.
3. If chalking returns abnormally fast the job requires refinishing.

Chipping

Cause

1. Usually caused by flying gravel or bumping of corners or edges.
2. Finishes which are brittle tend to chip easily.
3. Excess film thicknesses.
4. Insufficient adhesion of colour to primer or primer to metal.

Prevention

1. Use reliable materials according to directions.
2. Avoid excess film thicknesses.

Repair

1. Mild chipping is usually touched up with enamel using the point of a small brush to improve appearance and delay rusting.
2. Severe chipping requires complete refinishing operations.

Colour (off-shade)

Cause

1. Refinish colour not thoroughly stirred.
2. Original finish faded or off colour due to weathering.
3. Original finish does not match master standard supplied by motor company.
4. Wax coatings on the original finish are discoloured. This is a common condition on light colours.
5. Refinish colour not properly applied. This can be a result of poor or improper amount of thinner, wrong air pressure, spraying too wet or dry. These will greatly influence the final appearance of metallic colours in particular.
6. Insufficient number of coats.
7. Colour burned by hard wheel polishing.
8. Metamerism – colour match only under one type of light.

Prevention

1. Stir all colours thoroughly to incorporate all pigment.
2. Remove wax from adjacent sections with a wax and grease remover.
3. Light polishing of adjacent panels often helps to produce a colour match to newly painted section.
4. When applying metallics adjust spraying technique to effect colour match. Higher air pressure, greater distance between spray gun and panel, thin, dry coats give lighter shades. Lower air pressure and wetter coats give darker shades.
5. To prevent rings on spot jobs, overlap sufficiently on original finish. Use care to avoid cutting through when polishing.
6. A sure way to avoid colour mismatch is to spray a small metal panel with the materials to be used, and compare with the original finish before painting the job.
7. In severe cases, the new material should be tinted to match the old finish.

Cracking

This section includes shrinking and splitting (of undercoats).

Cause

1. Material not thoroughly stirred before use. If some of the undercoat has been previously used without stirring thoroughly, it may have upset the pigment to binder ratio to such an extent that the remaining material will always give trouble.
2. Surface not completely clean.
3. Insufficient flatting and feather-edging.
4. Oil or water in airline.
5. Insufficient thinning of undercoat.
6. Use of wrong or poor thinner.
7. Undercoat applied too heavily, especially first coat.
8. Blowing of air on wet film in attempt to speed up flash-off and drying.
9. Applying undercoat on a too hot or too cold surface.
10. Insufficient drying time allowed for between coats.
11. Use of too coarse a grade of wet and dry.

Prevention

1. Stir undercoat thoroughly each time before use.
2. Use wax and grease remover to clean surface before and after.
3. Flat well with the recommended grit paper. Feather-edge well back into old finish.
4. Keep airlines free of water and oil.
5. Thin undercoat fully with the proper good-quality thinner.
6. Apply undercoat in medium-thin coats, especially the first coat.
7. Avoid blowing air on wet undercoat.
8. Allow room, metal, and material to come to equal temperatures, preferably between 16 and 24 °C.
9. Allow each coat to dry thoroughly before applying succeeding coat.

Repair

Flat out cracked area, or wash off with proper thinner or solvent, and repaint.

Crazing

This section includes alligatoring, crow-footing and hair-lining (of colour coats).

Cause

Good-quality modern finishes properly applied rarely, if ever, fail by crazing. The above film defects, all related, and named to describe their appearance, are generally the result of the following:

1. Application of new finish over an old finish which has already crazed. The crazing in the old finish may not have been visible without close inspection using a magnifying glass.
2. Use of poor-quality, too brittle finishes.
3. Application of new finish over an old finish of new undercoat which is too soft.
4. Application of new finish over undercoats which are too thick or not thoroughly dried.
5. Application of excessively thick finish coats.
6. Mixing of unproven additives, such as gloss improvers, with the colour coat.
7. Insufficient stirring of materials to incorporate all the pigments.
8. Use of shellac sealers.
9. Use of poor quality or improper thinner.

Prevention

1. Always use highest-quality materials; they are much cheaper in the long run.
2. Inspect the old finish carefully with a magnifier. If old finish shows any signs of crazing, remove completely before refinishing.
3. Apply undercoats in medium-thin coats, allowing plenty of drying time for each coat.
4. Do not use too much material.
5. If the job has previously been repainted, flat

Paint defects: cause, prevention and repair 175

down thoroughly to reduce total film thickness.
6 Do not add anything to refinishing materials not specifically recommended by the manufacturer; then follow directions carefully.
7 Stir all materials thoroughly each time before use. Thinned materials settle more rapidly, so be sure to agitate material in spray cup occasionally during use.
8 Always use good and correct thinner, as well as a sufficient amount.

Repair

Thoroughly remove entire crazed finish and repaint.

Dirt

This section includes grittiness and seediness.

Cause

1 Particles flying about and settling on wet paint film. Particles may come from ordinary dirt and dust, dry sanding and grinding operations, fabric, clothing.
2 Dirt blown out of cracks, mouldings, under fenders, door jambs, etc., during spraying.
3 Failure to tack-off job properly immediately before spraying.
4 Improper stirring of colours which have become settled.
5 Failure to properly remove any skin formation on part-full containers, or poorly sealed containers of synthetic enamel.
6 Improper straining of colour.

Prevention

1 Use good housekeeping methods, keep shop clean. Don't spray in a room where any other work is being done.
2 Blow out all crevices, mouldings, remove dirt from underside of fenders before painting.
3 Seal door jambs and edges with a thin wet coat of the colour before painting surface.
4 Wetting down walls and floor with water may be necessary in some shops.
5 Tack off job thoroughly with new tack rags immediately prior to application of colour coats, including the masked areas.
6 Correct air pressure, and use a good thinner to avoid excessive spray dust.
7 Stir colour properly and thoroughly.
8 Thin to spraying consistency by adding the correct thinner.
9 Strain thinned material through finest strainer possible.
10 Do not poke or scrape strainer.

Repair

1 Dirt can sometimes be polished out with compound.
2 Films of enamel containing dirt should be allowed to dry thoroughly, the longer the better, before attempting to polish with fine compound or liquid polish.
3 Extremely dirty jobs should be allowed to harden thoroughly, wet flatted with P600 grit or finer paper, and resprayed.

Dry spray

See also *metallic faults*.

Cause

1 Poor-quality thinner (too fast).
2 Incorrect amount of thinner.
3 Excessive air pressure.
4 Improper spray gun setting, or dirty spray gun.
5 Holding spray gun too far from surface.
6 Spraying in draught.

Prevention

1 Use correct amount of thinner of the correct quality.

2. Use correct air pressure.
3. Adjust fluid feed and spray pattern to minimize overspray. Use clean and efficient spray gun.
4. Hold spray gun closer to work. Use good gun technique.
5. Do not spray in strong draughts.

Repair

In certain cases overspray can usually be removed by:

1. Polishing out.

With synthetic enamels it is important to prevent the overspray.

Drying

This section includes slow drying, wet spots and tack.

Cause

1. Painting over surface contaminated with wax, silicone, oil, fingermarks, petrolresidue or paint stripper.
2. Oil in airline.
3. Wrong type or poor-quality thinner.
4. Improper materials mixed with undercoat or colour.
5. Application of too much material.
6. Poor ventilation in drying area.
7. Atmosphere too humid or too cold.
8. Drier left out of synthetic enamel.

Prevention

1. Clean surfaces thoroughly with a wax and grease remover prior to painting.
2. Keep airlines free of oil. Keep compressor in good condition to prevent oil leakage into lines.
3. Use amount and type of thinner recommended by the paint manufacturer.
4. Do not mix anything with materials except as specifically recommended by the paint manufacturer.
5. Avoid piling on too heavy coats and too much material.
6. Make sure drying area maintains fresh air circulation.
7. Maintain steady temperature of drying room at about 21 °C to obtain satisfactory drying.

Repair

1. Wet spots are difficult to repair, and for good results they should be removed, thoroughly cleaned, and repainted.
2. General slow drying may be overcome by application of low heat. Use caution: if the condition is due to extra heavy coats, wrinkling may develop.

Fading (colour change)

Cause

Fading is a change in colour of the pigment portion of the film due to exposure to sunlight, or in a few cases to chemical fumes. Do not confuse fading with chalking or dulling. Fading generally goes deep into the film, and unlike chalking and dulling cannot usually be polished out.

Prevention

Some coloured pigments and pigment combinations are more resistant to fading than others. It is very difficult to prevent fading and about the only measures the painter can take to help are:

1. Use paint developed and made by a reliable manufacturer.
2. Use enough colour on the job to obtain good hiding.
3. Tint with only the bases recommended by the paint company for the type of colour in question.

Repair

If the job is badly faded (after it has been polished out) the only cure is to repaint.

Fish-eyes (silicone)

Cause

Surface contaminated with silicone, wax, grease or oil.

Many modern waxes and car polishes contain silicone, which is today the most common cause of fish-eyeing of fresh paint coatings. Silicones adhere strongly to the paint film, and extra effort is required for removal.

Minute quantities of silicone in flatting dust, from contaminated rags, or even from car polishes sprayed at considerable distances from jobs ready for painting can cause trouble.

Prevention

1. Clean surface thoroughly with wax and grease remover *before* flatting operations. Be sure to clean an area several times larger than repair area. Use plenty of wax and grease remover on clean rags. Do not allow to dry by itself, but quickly wipe dry with new, clean rags.
2. These rags should not be used again on any surface to be painted, since the slightest amount of silicone contamination will cause trouble.
3. Proceed with flatting operation, being careful not to touch or flat into uncleaned area.
4. Clean off flatting dust and do all blowing off with air at this point only.
5. Repeat wax and grease cleaning operation carefully. Treat bare metal areas with metal pretreatment and rust remover.
6. Tack off and start painting.
7. A simple test for silicone on a surface consists of spraying black or a dark coloured air-drying enamel one thin wet coat over a small area. Presence of silicone on surface will usually produce immediate formation of fish-eyes in the wet film. Wipe off with solvent cleaner before the test spot dries. This test is not infallible since the test section selected may not have any silicone on it for some reason.

Flatting scratches

This section includes lifting, raising or swelling of flatting scratches. This is often a mild form of lifting (see *lifting*).

Cause

1. Light scruffing or scratching of old surface, usually with too coarse grit wet and dry.
2. Insufficient flatting of old finish or undercoat.
3. Use of coarse grit wet and dry.
4. Improper cleaning of surface.
5. Flatting across area being repaired and old surface which contains silicone. This really brings the scratches up.
6. Failure to allow undercoat to dry thoroughly before flatting.
7. Flatting with petrol.
8. Spraying filler coats too heavily, especially the first coat over old finishes.
9. Use of poor-quality or wrong thinner.

Prevention

1. Use wax and grease remover and plenty of new clean rags to clean an area several times larger than repair area.
2. Use fine grit wet and dry, flatting well back into old finish.
3. Use recommended type and amount of thinner in primer–filler.
4. First coat of primer should be applied wet but thin.
5. Allow enough time between coats of filler for each one to dry.
6. Do not blow air on primer-filler to make it dry faster.

7 Allow surfaces to harden thoroughly before flatting.
8 Never use petrol for flatting old finish or fillers.
9 Flat with P600 grit wet and dry.
10 Allow to dry out thoroughly before applying colour.
11 Thin colour with amount and type of thinner recommended by paint manufacturer. Avoid cheap thinner.
12 Apply as recommended by the manufacturer.

Repair

1 Allow film to harden thoroughly, then rub out and polish.
2 If scratches are still too noticeable, wet flat with P600 grit paper and respray.

Flooding

This section includes floating, mottling and shadowing.

Cause

Most colours are made from a combination of different pigments which have varying densities and particle sizes, giving them a natural tendency to separate when the film is in a liquid state. Under normal conditions this tendency is small in magnitude and cannot be seen by the naked eye.

Certain conditions which may aggravate this to a point where the separation of the pigments become visible are:

1 Use of thinner which dries too slowly, allowing the pigment particles to migrate.
2 Applying the colour on a cold surface, or in a cold room.
3 Applying too heavy films of colour.

Prevention

1 Apply undercoats in normal recommended film thickness, and allow to dry thoroughly before application of colour coat.
2 Use thinner as recommended by the paint manufacturer.
3 Avoid extremely heavy coats.
4 Do not spray on a cold surface or in a cold room.
5 Iridescent colours may require a final light coat to avoid flooded appearance.

Repair

1 While film is still wet, apply a thin mist coat of colour.
2 If film is dry, respraying or refinishing is necessary.

Gloss

This section includes poor gloss, bloom, clouding, deadening and dulling.

Cause

1 Poor hold-out of undercoat.
2 Application of colour over surface contaminated with wax, grease, oil, soap or water.
3 Applying colour over undercoat which is not thoroughly hard and dry.
4 Use of poor-quality, too fast, or incorrect thinner.
5 Mixing improper additives with finishes.
6 Drying in closed room. Failure to provide sufficient circulation of fresh air.
7 Drying atmosphere extremely humid or cold.
8 Overspray (see *dry spray*).
9 Compounding or polishing colour coat too soon after application.
10 Using compound which is too coarse.
11 Use of strong detergents, soap, solvent or chemical cleaners on finish.
12 Insufficient film thickness of colour coat.
13 Blushing (see *blushing*).
14 Application of enamel over heavily chalked finish.

Prevention

1. Use undercoat having sufficient hold-out (see *hold-out*).
2. Use wax and grease remover, according to directions, over surface just prior to applying colour coats.
3. Allow undercoat to harden thoroughly before applying colour.
4. Always use the correct, good-quality thinner as recommended by the paint manufacturer.
5. Never add anything to the undercoat or the colour except as approved by the manufacturer of the paint.
6. Films should always have plenty of fresh, clean warm air circulation (not draughty) while drying.
7. Always stir all materials thoroughly each time before using.
8. Do not use compound or polish on a finish before it is thoroughly dry and hard. If a sensitive nose can detect *any* odour coming from the film, then it is too fresh to compound.
9. Use compounds and polishes free of coarse, gritty particles, oily vehicles, and strong solvents and chemicals.
10. Do not wash car with strong detergents, soap, solvents, or chemical cleaners. Use mild detergent and water.
11. Apply sufficient film thicknesses of colour as recommended.

Repair

1. After film is quite dry, gloss can usually be brought up by using a fine compound and light polish.
2. If polishing as above does not produce the desired gloss, refinishing is necessary.

Hiding

This section covers poor hiding, poor covering, transparent film and undercoat showing through.

Cause

1. Colour not thoroughly stirred.
2. Use of too slow thinner, thereby preventing application of sufficient film thickness.
3. Not enough material used.
4. Inadequate film thickness.
5. Too much polishing of colour coat.

Prevention

1. Stir all colours thoroughly.
2. Use thinner recommended by the paint manufacturer.
3. Use normal number of coats.
4. Spray under good lighting conditions to get uniform coating over entire job.
5. Avoid over compounding or polishing. Be especially careful on edges and corners, as it is very easy to remove too much of the colour.

Hold-out

This section covers poor hold-out, porosity and sinking in.

Cause

1. Insufficient stirring of undercoat *each time* before use.
2. Use of poor-quality or insufficient thinner in the filler coats.
3. Spraying of fillers too dry, resulting in porosity.
4. Use of too much fillers.
5. Applying colour coat before filler is thoroughly dry.
6. Flatting filler before it is thoroughly hard.
7. Using too coarse grit wet and dry.
8. Insufficient flatting of filler.
9. Insufficient amount of colour coat.

Prevention

1. Do not use excessive thinner in enamel colours.

2. Use thinner as specified by the manufacturer of the filler and colour.
3. Stir filler thoroughly *each time* before use.
4. Spray fillers as instructions.
5. Allow each coat of filler to dry before applying next one.
6. Allow filler to become thoroughly hard and dry before flatting and apply colour coat.
7. Use fine grit wet or dry. Final flatting should be done with grit not coarser than P400, P600 grit paper which will give much better hold-out.
8. Do not use too much filler. Flat down thoroughly.
9. Use sufficient amount of colour coat.

Repair

1. When colour coat is completely hardened (several days may sometimes be required) wash with water, and rub out with fine compound and liquid polish.
2. If polishing as above does not give satisfactory appearance, refinishing is required.

Lifting

This section includes raising and crinkling (see also *wrinkling*).

Cause

1. Lifting is caused by solvents in a refinish paint attacking a previously painted surface. If the previous film is completely soluble in the wet coating applied over it, or if the film is completely impermeable (won't let it soak in) to the wet coating applied over it, lifting will *not* occur. Lifting occurs in varying degrees from the mild forms of swelling or raising of flatting scratches to severe lifting and destruction of the base film. All fast dry synthetic paint films have a certain range of time after application when, under certain conditions, they can be made to lift upon recoating. Paint manufacturers' instructions and recommendations are planned to avoid these time ranges and conditions.
2. Second coat is more apt to produce lifting if first coat is applied over surface not thoroughly cleaned and flatted.
3. Use of cellulose over air synthetic enamel.
4. Application of second coat of fast dry synthetic over the first coat which has become partially dried.
5. Use of cellulose thinner in enamel.
6. Too heavy applications of synthetic enamel.
7. Use of too coarse grit wet and dry paper.
8. Application of colour over incompletely cured undercoat.

Prevention

1. Begin with a clean, well-flatted surface.
2. Do not use cellulose thinner in synthetic enamels.
3. Avoid use of cellulose over synthetic enamel.
4. Second coats of fast dry synthetic should be sprayed over first coat immediately after application of first coat, or after first coat has completely dried.
5. Avoid heavy coats.
6. Allow each coat to flash off before following coat is applied.
7. Don't apply colour over undercoat which is not completely dry and hard.

Repair

The only recommended method of repair is to completely remove the lifted area and refinish (see also *scratches* and *wrinkling*).

Metallic faults

Metallic faults, i.e. banding, ringing etc., are usually caused by poor gun techniques.

Banding

This section covers stripes in panel and tram lines.

Cause

1. Poor overlap of gun on each pass.
2. Gun too near panel.
3. Thinner too slow.
4. Gun not held at right angles to panel.

Prevention

1. Overlap each pass of gun by 50 per cent.
2. Hold gun 150–250 mm (6–10 in) from panel.
3. Use suitable thinner, i.e. faster.
4. Hold gun at right angles to panel. Special care on bonnet, roof etc.

Repair

Apply another coat, observing above.

Ringing

This section includes flotation and very wet coat.

Cause

1. Coats too wet.
2. Gun held too close.
3. Fluid control open too wide.
4. Gun moved too slowly.
5. Air pressure too low.

Prevention

1. Hold gun 150–250 mm (6–10 in) from panel.
2. Close fluid control.
3. Move gun faster.
4. Raise air pressure.

Repair

Apply another coat after flash-off.

Dry coats

Cause

1. Gun held too far from panel.
2. Fluid control closed.
3. Gun moved too fast.
4. Air pressure too high.

Prevention

1. Hold gun 150–250 mm (6–10 in) from panel.
2. Open fluid control.
3. Move gun slower.
4. Lower air pressure.

Repair

1. Apply another coat.
2. If very dry, flat with P600 and apply colour coat.

Orange peel

This section covers poor flow and pebbling.

Cause

1. Spraying over surface contaminated with wax, grease and especially silicone.
2. Using wrong type or poor grade thinner.
3. Insufficient thinning of colour.
4. Too high air pressure.
5. Improper adjustment of spray gun.
6. Poor spray gun technique. Holding gun too far from or too close to surface.
7. Spraying in draught.
8. Coats applied too dry and thin.
9. Cold shop or metal temperatures.
10. Plugged up spray gun or airline.

Prevention

1. Spray over properly cleaned surface, completely free of wax, oil and silicone.
2. Use thinner as recommended by paint manufacturer. Use a little *more* thinner.
3. Use lowest air pressure that will give good atomization.
4. Clean and adjust spray gun properly.
5. Use good spray gun technique. Hold spray

gun 150–250 mm (6–10 in) from surface, and always keep gun at right angles to the area being sprayed.
6. Prevent draughts on job.
7. Apply wet coats of thin to medium thicknesses.
8. Shop and metal should be at normal temperatures, preferably between 16 and 24 °C.
9. Compressor should supply sufficient air. Be sure air transformer and lines are not dirty or plugged up.

Repair

1. After colour is thoroughly hardened, rub out the orange peel with a fine compound and polish.
2. If condition is very bad, wet flat with P600 wet and dry and respray.

Pinholing

This section includes pitting and pock marks.

Pinholding and pitting are generally caused by the same conditions that produce blistering. Pinholes and pits are tiny blisters which have broken open at the surface during the drying stage, or blisters which have been broken open by rubbing and polishing the dried film (see *blistering*).

Precipitation

This section includes break, compatibility, curdling and throw-out.

Cause

1. Dumping in, or addition of, thinner too rapidly to paint.
2. Insufficient stirring.
3. Use of wrong type of thinner.
4. Use of old paint which has become partially oxidized and insoluble.

Prevention

1. Add thinner to the paint slowly while constantly stirring.
2. Do not use old paint which shows excessive skinning.
3. Add correct amount of thinner.

Runs

This section includes curtains and sags.

Cause

1. Spraying over surface contaminated with wax, oil, grease or silicone.
2. Use of too much slow-drying thinner.
3. Coats applied too heavy and wet.
4. Poor spray gun technique or adjustment. Distorted spray pattern.
5. Material, surface or atmosphere too cold.
6. Air pressure excessively low.

Prevention

1. Clean surface thoroughly with wax and grease remover before painting.
2. Avoid use of thinner which is too slow drying. Follow paint manufacturer's instructions as to type and amount of thinner.
3. Do not apply heavy coats. Rely on thin to medium coats.
4. Clean and adjust spray gun to give proper, uniform spray pattern. Develop and practice good spray gun technique. Use good triggering action. Curved surfaces require extra careful gun action.
5. Since different brands of paint and thinner require somewhat different techniques, practice on old panel when switching to new materials to obtain the 'feel' of the material.
6. Adjust air pressure and spray gun settings and technique to compensate for temperature and weather conditions.
7. Be sure spray booth allows plenty of room for easy working in uncramped position.

8 Spraying should never be done without plenty of properly located light. Many runs are the result of not being able to see the area being sprayed.

Repair

1 If sag is still wet, and another coat can be applied, lightly brush out the sag with a fine camel hair brush, then recoat.
2 When colour coat is completely dry and hard, flat out sag with P600 grit wet and dry. Polish out or respray as necessary.

Rusting

Cause

1 Moisture and chemicals (such as road salt, etc.) attacking the metal through visible or microscopic breaks in the paint film. Rusting then often works back under the film, resulting in blistering and peeling. Rusting also forms more rapidly at points where the moisture collects and remains in contact with the film.
2 Painting over metal which contains rust not completely removed will always result in rapid failure from more rusting.
3 Painting over metal touched by bare hands, or metal contaminated by chemical deposits from flatting water.

Prevention

1 Bare metal should be flatted very thoroughly to remove all traces of rust from surface and pits in the metal.
2 Metal should be treated with metal pretreatment and rust remover according to directions.
3 Bare metal areas should never be touched by the hands after using metal pretreatment and rust remover and should be primed within 30 minutes to prevent start of new rust formation.

Repair

1 Remove paint down to metal.
2 Repaint following steps under 'Prevention'.

Settling

This section includes hard settling and caking.

Cause

All pigmented materials will settle. Many factors, such as type of pigment, formulation, fineness of grind, and viscosity will determine whether the pigments settle hard, soft, fast or slow.

Prevention

Steps which can be taken by the paint dealer and the paint user to minimize settling are:

1 Never store paint near heat. Keep in a cool place.
2 Invert containers periodically.
3 Use oldest stock first.
4 Never store thinned paint. Never pour paint back into container of unthinned paint.

Repair

1 All paint products should be stirred thoroughly each time before use. Materials which do not appear to be settled may actually contain a much higher concentration of pigment near the bottom of the container than near the top, and still be so soft that the settling cannot be detected. Thus an undercoat, if used without stirring, may give poor adhesion, hard flatting, etc.
2 If firm or hard settling is present, the liquid portion should all be poured off into a separate clean container, and the settled portion should be broken up and stirred thoroughly until completely smooth and free of lumps and visible particles.

The liquid portion should then be poured

back into the original container, very slowly at first, while continually stirring the material.
3 After thinning, the material should be strained through a fine straining medium.

Skinning

Cause

1 Drying and oxidation of the liquid surface of a paint, particularly in reference to synthetics, forming a soft and slimy or dry and tough film at the surface.
2 Storing of synthetics in a partly filled container.
3 Storing of synthetics in a container which is not closed airtight.
4 Opening and reclosing containers of synthetics a number of times.

Prevention

1 Keep containers tightly sealed.
2 Do not save small amounts of synthetics.

Repair

1 Carefully remove all skin formation *before* stirring or shaking material. Failure to do this may result in bitty or seedy finish.
2 If skin is excessively thick, material should be discarded and not used.

Staining

This section includes spotting (see also *water spotting*) and industrial fallout.

Cause

1 Oil in air line.
2 Use of poor quality wax or polish.
3 Contamination of surface by chemically reactive materials such as cement dust, tree sap, acid droplets, antifreeze solutions, bleaching powders, bird droppings, wet leaves, road oil and tar droplets, insects, ink, antiseptics, industrial fumes, etc.
4 Paint droplets settling on finish.

Prevention

1 Keep compressor in good condition to avoid oil pumping.
2 Check and drain air transformer at least once each day.
3 Check compressed air for oil before spraying.
4 If possible, avoid parking car under trees, or in localities where chemical dusts and fumes, and droplets from industrial processes can get on finish.
5 Wash off any foreign material that gets on finish immediately.
6 Do not use unproven types of waxes and polishes.

Repair

1 Wash off with clear water or mild detergent solution in water.
2 Rub spots lightly with wax and grease remover.
3 Use a fine rubbing compound, and liquid polish.
4 After above treatment allow finish to weather a few days.
5 If the industrial fallout is identified as being principally iron in nature, for example from foundry dust and from chimneys etc., such deposits may be removed by adopting the oxadic acid wash procedure, providing that action is taken at an early stage (see Chapter 8).
6 If this treatment does not give satisfaction refinishing is required.

Streaking

This section includes lapping marks.

Cause

1. Distorted spray pattern. Spray gun dirty or needing repair.
2. Poor spraying technique. Insufficient overlap when spraying.
3. Spraying metallic colours too wet. Spraying any colour too dry.
4. Colour insufficiently thinned.
5. Use of wrong thinner.
6. Metal too hot or too cold.

Prevention

1. Check spray pattern each time before starting job. If not uniform, clean spray gun thoroughly. If pattern is still distorted make necessary repairs to spray gun.
2. Use correct spray technique, with sufficient overlapping of strokes.
3. Be sure colour is thinned with the proper amount and type of thinner.
4. Allow metal to come to normal room temperature before spraying.

Repair

1. While film is still wet, recoat with thin mist coat to eliminate streaky appearance in metallic colours.
2. Allow to dry completely and recoat.

Water spotting

This section includes rain spotting.

Cause

1. Application of excessively thick coats.
2. Allowing rain or dew to get on finish, or washing finish before it is thoroughly dry. Combination of rain and sunshine on fresh paint job is particularly likely to cause water spotting.
3. Tap water as used in car washing or lawn sprinkling may contain minerals that if allowed to remain on a finish may cause spotting.
4. Boiling water from overheated radiator getting on finish.
5. Frequently more noticeable on certain colours and on cars which have had excessive wax application.

Prevention

1. Apply colour in normal film thickness.
2. Do not allow water to get on finish until it is thoroughly dry.
3. Do not wash car in bright sunshine allowing water droplets to remain on surface. Water drops act like small burning lenses, magnifying the heat from the sun.
4. Avoid overuse of waxes.

Repair

1. Wash off with cool water. Rub lightly with damp rag, then polish out.
2. If polishing does not give satisfactory results, area must be repainted.

Wrinkling

This section includes crinkling and puckering.

Cause

1. Rapid drying of top surface while the underneath remains soft.
2. Any condition that leads to thick films will tend to cause wrinkling: see *runs*.
3. Any condition that produces a lifting tendency will often initiate or aggravate wrinkling: see *lifting*.
4. Many conditions that produce slow drying tend to produce wrinkling: see *drying*.
5. Spraying in hot sun, or exposing to sunshine before synthetic paint is thoroughly dry.
6. Abnormally hot and humid weather.

Prevention

1 Avoid all conditions which may lead to *runs*, *lifting*, and slow *drying*.
2 At all costs avoid application of abnormally thick films.
3 If possible avoid spraying in abnormally hot humid weather.
4 Do not spray in hot sun, or run car into sunshine before it is completely dry.
5 Do not put water on freshly painted enamel job, especially if surface is warm.

Repair

1 The only recommended course of action is to remove the wrinkled film and repaint properly.

11 Multiple-choice questions

Multichoice assessment has, over the years, proved to be a most popular method of measuring 'knowledge of subject matter'. There are, for one thing, economies for the examining body. As far as the student is concerned, perhaps the most important consideration is that there is no need to worry so much about how to organize the material (let alone spell it) in response to the questioning. There are, however, many alternatives for the student confronted with multichoice answers. The solution may be known and underlined immediately; an educated guess may be made (or even a blind guess); the very obviously incorrect answers may be eliminated; or one of the golden rules of mathematics may be applied, that is, 'If it looks wrong, it is wrong; if it looks as if it might be right, the chances are it's still wrong'.

Owing to the stress imposed by examination conditions and under-preparation, many students have admitted to thinking that the examiner is trying to trick them. In view of the nature and arrangement of some of the questions and solutions this may appear to be true but, of course, it is not, and the idea should be dismissed once and for all. The student should not experience too much difficulty, if any at all, if much in-course multichoice examination practice is undertaken. It is quick and easy to mark from the teacher's point of view and builds confidence for the student.

The following are typical examples of multichoice questions, relating first to light vehicle body repair and then to a combination of paint spraying work and paintwork defects and correction. The correct answers can be found on page 198.

Attempt all questions. Read each question carefully. Select the correct answer from those given and place a tick against it. There are 100 questions.

1. To align the inner wing panels of a motor car with the strut type of suspension it is necessary to have
   (a) cramps
   (b) pull hydraulics
   (c) front-end jigs
   (d) push hydraulics

2. The purpose of spring-hammering with the use of a body spoon is to
   (a) stiffen flat panels
   (b) prise out dents between panels
   (c) reduce ridges on panels
   (d) shrink damaged areas

3. Hydraulic pushing equipment should not be used on a vehicle body which is
   (a) twisted
   (b) corroded
   (c) misaligned
   (d) distorted

4. A motor body of integral construction may be identified by the
   (a) absence of a separate chassis
   (b) wooden framework clad in aluminium
   (c) inclusion of a separate chassis
   (d) wooden framework reinforced with gussets

5 A beating file is used instead of a hammer for dressing out a panel in order to
(a) raise the surface
(b) rough out the surface
(c) reduce the stretching
(d) round out the surface

6 A hide mallet is mainly used on sheet metal to
(a) planish the surface
(b) protect the surface
(c) prevent surface stretching
(d) avoid surface shrinking

7 The most likely effect of hollowing out sheet metal would be to
(a) produce cracks at the outer edges
(b) thin out the metal at the central area
(c) maintain the thickness throughout the form
(d) thicken up the central area of the form

8 The action needed if an oxy-acetylene welding pipe were accidentally saturated in oil would be to
(a) clean it with a dry rag
(b) wash it in soapy water
(c) replace it immediately
(d) clean it with a solvent

9 Which one of the following is a structural component?
(a) front wing
(b) door
(c) bonnet panel
(d) cant rail

10 The technique of indirect hammering involves hitting the metal
(a) with excessive force
(b) off-the-dolly
(c) on-the-dolly
(d) with the edge of the dolly

11 A misaligned door can be rectified by adjusting the
(a) dovetails
(b) hinges
(c) striking plate
(d) lock mechanism

12 When a dented front wing is being repaired, the bitumen-based underseal should be
(a) 'sanded off' with a disc sander
(b) heated with a welding torch and rubbed clean with a wire brush
(c) removed with an electric rotary wire brush
(d) scraped off with a scraper and wiped clean with a rag soaked in paraffin or turpentine

13 As a general guide, the sequence of operations during the repair of a collision damaged motor car should be
(a) dictated by the amount of heat treatment involved in repairing the damage
(b) the reverse of the sequence of damage sustained during collision
(c) dictated by the amount of resistance welding involved in repairing the damage
(d) the same as the sequence of damage sustained during collision

14 Fluxes are used during the soft soldering process to
(a) prevent or reduce heat transfer
(b) prevent the formation of surface oxides
(c) remove oil and grease
(d) lower the melting point of the solder

15 The main function of the aperture made during manufacture in a car inner door panel is to
(a) provide location in the press on production
(b) give access to the window winding mechanism
(c) provide adequate ventilation
(d) decrease the weight of the door

16 Compressed air lines in workshops should be colour coded to
(a) indicate their contents
(b) reduce heat loss
(c) prevent corrosion
(d) blend in with the surroundings

17 Locked-in stresses may occur as a result of gas welding steel body panels and may be relieved by
(a) shrinking
(b) planishing
(c) grinding
(d) tempering

18 The technique of direct hammering on a dolly is used to
(a) planish
(b) rough out
(c) hot shrink
(d) hollow

19 A swaging machine shapes the metal by
(a) matching male and female rollers
(b) vibratory tools
(c) a pressing action
(d) offset wheels

20 Shutting down the oxy-acetylene plant after welding calls for the following actions:
1 close the oxygen valve at the torch
2 close the acetylene valve at the torch
3 release the regulator pressures
4 close the cylinder (bottle) valves
The correct sequence is
(a) 2 1 3 4
(b) 2 1 4 3
(c) 4 1 2 3
(d) 4 2 1 3

21 In metal arc welding the use of an excessively high current setting would result in the weld zone having defects of
(a) intermittent weld bead deposit
(b) poor penetration and bead overlap
(c) excessive spatter and flat bead
(d) porous weld bead and spatter

22 Swaging is used to
(a) flange metal sections
(b) stiffen panels
(c) produce double-curvature shapes
(d) straighten panels

23 Which one of the following sheet metal processes will produce a high crown surface?
(a) wheeling
(b) puckering
(c) planishing
(d) hollowing

24 The major stress encountered by a vehicle body of monocoque (integral) construction is
(a) compressive
(b) torsional
(c) tensile
(d) shear

25 When a jacked-up vehicle is being worked on, it should be
(a) wedged with wood under the wheels
(b) left free for easy access
(c) supported by axle stands
(d) left on the jack

26 Protection by the anodizing process is based upon the principle that
(a) the hot dipping of aluminium requires only low temperatures
(b) the natural oxide film on aluminium can be artificially thickened
(c) aluminium is one of the easiest metals to electroplate
(d) there is no limit to the size of the article treated

27 The chief characteristic of a body solder comprising 70 per cent lead and 30 per cent tin is
(a) low melting point
(b) adhesive properties
(c) wide pasty range
(d) high melting point

28 Which one of the following must be present for corrosion to take place by electrochemical action?
(a) paint
(b) water
(c) dirt
(d) gas

29 The adhesive which resists heat, water and acid, and is used to join metals, is made from
(a) ureaformaldehyde
(b) epoxy resin
(c) phenol formaldehyde
(d) polyester resin

30 Fuses are included in electrical circuits to
(a) avoid the need for earthing each component
(b) reduce the power factor
(c) limit the current in the circuit
(d) isolate the machine during repair

31 The maximum hourly draw-off rate for an acetylene gas cylinder is
(a) one-quarter of its contents
(b) one-fifth of its contents
(c) one-third of its contents
(d) one-half of its contents

32 What is the name given to a group of steels containing a minimum of 9% chromium?
(a) stainless steel
(b) nickel steel
(c) medium-carbon steel
(d) low-carbon steel

33 Accident damage to a glass-fibre reinforced plastic body is often easy to repair because
(a) it requires no skill or knowledge to use GRP
(b) the body will not rust if the paint film is cracked
(c) damage is limited to the point of impact
(d) the repaired area will not need painting

34 The main purpose of a body jig is that it
(a) allows greater pressures to be used when pulling
(b) facilitates alignment of the underbody
(c) allows less material to be used in the vehicle body
(d) gives better conduction of heat when straightening

35 When the manufacture of a glass-fibre reinforced plastic laminate is complete, its properties cannot be altered by further heat treatment because
(a) plastics materials do not permit the use of heat treatment
(b) the laminate was irreversibly cured by the action of a catalyst
(c) the risk of fire would be too great
(d) the material would melt if subjected to heat

36 Which one of the following combinations identifies acetylene gas?
(a) distinct odour, combustible, left-hand thread, maroon cylinder
(b) left-hand thread, distinct odour, red cylinder, brass regulator
(c) combustible, maroon cylinder, right-hand thread, odourless
(d) maroon cylinder, left-hand thread, distinct odour, non-combustible.

37 Which one of the following soldering fluxes is an active or corrosive flux?
(a) tallow   (c) petroleum jelly
(b) zinc chloride (d) resin

38 One workshop method of determining the correct temperature of aluminium during annealing is
(a) using a thermometer attached to the surface
(b) heating the metal for a set period of time
(c) touching the surface with a dry piece of wood until it leaves a mark
(d) watching the surface for colour change

39 A fundamental difference between a riveted joint and a high-duty adhesive-bonded joint is the
(a) failure of the bonded joint in shear
(b) relatively small surface area occupied by the rivets
(c) electrolytic potential of the riveted joint
(d) failure of the bonded joint in water

40 One important advantage of using a welding jig for vehicle body components is the
(a) variation in the type of article to be welded
(b) free contraction of the components
(c) self-location of components to be welded
(d) use of lower fusion temperatures

41 Pneumatic hand power tools may be used in preference to electrical hand power tools because
(a) work can still continue during a long power cut
(b) they are safe in damp or humid conditions
(c) the air hose can be extended considerably without loss of power
(d) compressed air is always available in the workshop.

42 An aluminium die cast lamp housing has to be fitted to a vehicle panel. An insulator would be required if the panel was made from
(a) aluminium
(b) low-carbon steel
(c) glass-fibre reinforced plastic
(d) plywood

43 The list below shows a number of metal/cutting fluid combinations:
1 aluminium–soluble oil
2 steel–paraffin
3 brass–dry (no fluid)
4 aluminium–paraffin
5 steel–dry (no fluid)
6 brass–soluble oil
7 aluminium–dry (no fluid)
8 steel–soluble oil
The correct combinations are given in lines
(a) 1, 5 and 6
(b) 4, 3 and 8
(c) 3, 5 and 1
(d) 2, 4 and 6

44 The back stops fitted to power-operated guillotines are used to
(a) limit the amount of scrap
(b) prevent cut sheet metal from falling
(c) enable repetition cuts to all given dimensions
(d) provide a safety gate

45 The dangerous fumes given off when welding galvanized sheet metal are
(a) zinc oxide
(b) sulphur dioxide
(c) carbon monoxide
(d) trichloroethylene

46 A material that has the ability to be drawn into a thin wire is said to be
(a) tough   (c) malleable
(b) hard   (d) ductile

47 The principal grades of files in order of coarseness are
(a) dead smooth, smooth, second cut, bastard, rough
(b) dead smooth, smooth, second cut, rough, bastard
(c) smooth, dead smooth, second cut, rough bastard
(d) dead smooth, second cut, smooth, bastard, rough

48 The main purpose of the chassis legs on an integrally constructed vehicle is to
(a) carry mechanical components
(b) strengthen the lower bulkhead
(c) give a rigid protection in an accident
(d) give added strength to the whole vehicle

49 By law, checks are periodically made on air compressors to determine that they are operating at their
   (a) correct speed
   (b) safe working load
   (c) correct temperature
   (d) safe working pressure

50 A flashback arrestor is fitted into an oxy-acetylene plant between the
   (a) oxygen and acetylene regulators
   (b) regulator and cylinders
   (c) regulator and torch
   (d) welding tip and mixing chamber

51 For laying up a glass-fibre reinforced plastics laminate, resin drainage on vertical surfaces can be controlled by
   (a) adding glass fibres to the moulding
   (b) adding thixotropic agent to the formulation
   (c) increasing the amount of colouring pigment
   (d) an increased percentage of catalyst

52 A damaged door has to have a new outer skin fitted. Which of the following pieces of equipment should be used to remove the edge of the damaged outer skin?
   (a) setting wrench
   (b) rotary sanding disc
   (c) flange tool
   (d) panel puller

53 An optical track gauge would be used for the alignment of
   (a) a chassis
   (b) steering
   (c) an integral body
   (d) window and door apertures

54 When applying body solder which one of the following operations is carried out first?
   (a) brushing flux
   (b) filling solder
   (c) heating the metal
   (d) tinning

55 Which one of the following conditions would be most likely to cause a porous weld when using MIG (metal inert gas) welding equipment to repair a motor car wing?
   (a) gauge of metal incompatible
   (b) welding gun moving too slowly
   (c) excessive shield gas pressure
   (d) working in draughty workshop

56 The most important function of protective clothing for general workshop use is to
   (a) present a tidy appearance
   (b) provide safe working dress
   (c) protect everyday clothes from damage
   (d) prevent everyday clothes from becoming dirty

57 The recommended method of welding when replacing part panels for minimum distortion is
   (a) resistance   (c) metal arc
   (b) MIG   (d) oxy-acetylene

58 An essential characteristic of body waterproofing compounds is that they should
   (a) be water solvent
   (b) never harden
   (c) set hard
   (d) be acid resistant

59 The use and purpose of a viscosity or flow cup is
   (a) to ensure thorough mixing of the solvent/paint mix
   (b) to ensure the correct paint/solvent ratio
   (c) to ensure that the paint mix is clean and filtered
   (d) to measure the quantity of paint required for a given area.

60 Paint manufacturers specify a flash-off time between coats which applies to average shop conditions. Which of the following conditions would make necessary a longer flash-off than that normally recommended?
   (a) cold, damp conditions

(b) high air movement
(c) hot, dry conditions
(d) the use of a fast thinner

61 The term 'spraying set-up' refers to a combination of three items which can be called the principal parts of the spray gun. These parts are
(a) fluid tip, fluid needle, air valve
(b) fluid tip, air valve, air cap
(c) fluid tip, fluid needle, air cap
(d) air valve, air cap, fluid tip

62 The air pressure at the spray gun will be
(a) the same as indicated by the air supply tank gauge
(b) as indicated by the transformer/regulator gauge
(c) higher than either (a) or (b)
(d) lower than either (a) or (b)

63 Which of the following conditions would *not* cause an air leak at the front of the spray gun?
(a) lumps of paint or dirt lodged in the fluid tip
(b) worn or damaged air valve seat
(c) bent air valve stem
(d) foreign matter on the air valve or seat

64 Which of the following conditions would *not* cause a jerky or fluttering spray pattern?
(a) loose fluid tip or damaged tip seat
(b) dry packing or loose fluid needle packing nut
(c) loose, dirty or damaged coupling nut or cup lid
(d) worn or damaged air valve or seat

65 A test spray pattern is seen to be bottom heavy, and does not alter when the air cap is turned through 180 degrees. The defective pattern is probably caused by
(a) blocked air cap horn holes
(b) paint too heavy for suction feed
(c) dried paint just inside the fluid tip
(d) dry packing or loose fluid needle packing nut

66 True low-bake enamels dry out and harden
(a) entirely by evaporation of the solvent content
(b) with initial dry by solvent evaporation and final hardening due to a chemical change in the vehicle or binder owing to the take-up of oxygen
(c) by chemical reaction between solvent and binder
(d) with initial dry by solvent evaporation and final hardening due to a chemical change in the vehicle owing to heat

67 A primer paint is designed to
(a) provide a paint coat with good scratch-filling characteristics
(b) provide a paint coat with good adhesive properties
(c) provide a paint coat which is easy to sand or flat
(d) provide a paint coat which will improve the finish coat gloss

68 When carrying out a spray pattern test with the paint correctly mixed, atomization is seen to be too fine (recognized by excessive overspray and dry spray). This can be corrected by
(a) reducing the air pressure
(b) increasing the air pressure
(c) reducing the material flow
(d) by adjusting the spreader valve

69 Which of the following statements concerning the spray gun spreader adjustment valve is true?
(a) it controls the limit of travel on the fluid needle
(b) it withdraws the needle from the fluid tip
(c) it controls the air supply to the air cap horn holes
(d) it lets more or less paint through the fluid tip nozzle

70 A primer–surfacer paint is designed to have good adhesive qualities like a primer but at the same time to be capable of filling in minor scratches. This additional quality of scratch filling is obtained by
(a) employing a special binder or resin
(b) the addition of fast solvents
(c) a much larger amount of pigment
(d) using less solvent in the mix

71 Why is it particularly important to be aware of the air pressure and the pressure differential between the transformer/regulator and the spray gun when spraying metallic finishes?
(a) because these finishes should be applied at a lower viscosity than 'straight' colours
(b) because these finishes require a different spray gun set-up
(c) because variations in the atomizing air pressure can vary the colour effect of these finishes
(d) because these finishes dry faster

72 If the spray gun is held at a greater distance from the work than that specified, the result will be
(a) a wet non-uniform coat with a tendency for sags or runs to develop
(b) a dry sandy uneven coat of paint with an increase in overspray
(c) a non-uniform deposit of paint giving an uneven spray pattern
(d) an increase in overspray with a tendency for sags and runs developing at the end of the stroke

73 If the spray gun fluid tip extends slightly beyond the air cap, the gun may be recognized as
(a) a suction feed gun
(b) a gun with an internal mix air cap
(c) a pressure feed gun
(d) a high-production gun

74 By adding a resin/thinner solution commonly known as 'hardener/thinner', it is possible to convert a type of paint which is normally air drying into a low-bake enamel. The type of paint which can be handled in this manner is
(a) an air-drying synthetic enamel
(b) an air-drying nitrocellulose lacquer
(c) an air-drying acrylic lacquer
(d) a two-pack polyurethane finish

75 What is the minimum temperature required to bake true low-bake enamels?
(a) 71–74 °C
(c) 82–85 °C
(c) 60–63 °C
(d) 65–68 °C

76 The recommended minimum sanded film thickness for the primer–surfacer coat is
(a) 0.0381 mm (0.0015 in)
(b) 0.0508 mm (0.002 in)
(c) 0.0635 mm (0.0025 in)
(d) 0.0762 mm (0.003 in)

77 The recommended air pressure for spraying acrylic lacquers is lower than that recommended to spray nitrocellulose lacquers and synthetic enamels. Why is this?
(a) because acrylic lacquers have a rapid initial surface dry
(b) because acrylic lacquers have a lower application viscosity
(c) because a different spraying set-up is used
(d) because the solid content of acrylic lacquers is higher than that of nitrocellulose lacquer

78 For efficient operation, extractor fans or extractor ducting inlets should be placed
(a) approximately 457.2 mm (18 in) to 609.6 mm (24 in) high
(b) as high as possible
(c) at least 1828.8 mm (72 in) to 2438.4 mm (96 in) high
(d) their position is immaterial

79 A fast thinner used in cold, damp, humid conditions can give rise to which of the following defects?
  (a) colour sinkage
  (b) striping
  (c) slow drying
  (d) blushing or blooming

80 Why is it recommended procedure to sponge off and leather dry after sanding or flatting a primer–surfacer coat?
  (a) to prevent the defect known as 'water spotting' occurring later
  (b) because water which is allowed to evaporate naturally can later cause blistering
  (c) to avoid abrasive scratches showing through the finish colour
  (d) to provide a tooth or key essential for good intercoat adhesion

81 Which of the following defects can occur during drying if heat is applied too soon to a wet paint film?
  (a) blisters
  (b) flaking
  (c) boil
  (d) blushing or blooming

82 The recommended method to remove severe 'industrial fallout' contamination is
  (a) to polish out the defect with compound
  (b) to use an oxalic acid solution of 8 parts crystals to 20 parts water
  (c) to use an oxalic acid solution of 2 parts of crystals to 20 parts water
  (d) to flat and respray affected areas

83 Which of the following defects will not respond at any time to a polish repair?
  (a) dry spray
  (b) orange peel
  (c) dirt in the paint
  (d) blisters

84 Among other reasons, stripping on metallic finishes can occur if
  (a) a wet coat is applied
  (b) insufficient drying time is allowed between coats
  (c) the spray gun is held too close to the surface
  (d) the material is not thoroughly stirred

85 If the local primer–surfacer coats applied during a 'patch to bare metal' repair are not completely dry before the colour coats are applied, which of the following defects is most likely to occur when the colour coat hardens?
  (a) off-colour
  (b) blushing or blooming
  (c) sinkage or contour mapping
  (d) water spotting

86 Which of the following defects requires a 'strip to bare metal' for rectification?
  (a) sags or runs
  (b) orange peel
  (c) lifting
  (d) water spotting

87 The application of a solid wax polish to a newly applied lacquer paint film can cause defects if applied too soon. The recommended time lapse before applying solid wax is
  (a) three to four weeks
  (b) one to two weeks
  (c) one to two days
  (d) three to four days

88 'Invariably due to faulty adhesion caused by inadequate preparation of the underlying surface, contamination, inadequate flatting, or failure to use metal treatment fluid on bare metals.' The foregoing is the likely cause of which of the following defects:
  (a) boil
  (b) orange peel
  (c) flaking
  (d) pits

89 A guide coat is used to
   (a) provide a film with good adhesion for the colour coats
   (b) to provide a coat which will improve hold-out and improve the finish coat gloss
   (c) to provide a coat of uniform colour so that a solid colour effect can be obtained with the minimum number of finish coats
   (d) to ensure that the minimum amount of surface is removed during sanding and highlight imperfections or hollows by guide coat which remains in these locations

90 Which of the following is true? The aluminium flakes in a metallic paint will be trapped in the surface of the film, to lighten and emphasize the metallic appearance,
   (a) if the material is sprayed dry
   (b) if full wet coats are applied
   (c) if the material is applied in full wet coats in cold conditions and slight air movement
   (d) if the spray gun is correctly adjusted but moved slowly across the surface

91 Dry spraying the undercoats, resulting in a porous undercoat film, can result in
   (a) flotation
   (b) slow drying
   (c) colour sinkage
   (d) scratch opening

92 When applying lacquer finish colour coats, initial film build-up should be obtained by
   (a) spraying a double coat to obtain good build, flow and gloss
   (b) applying double coats at high viscosity
   (c) applying single coats
   (d) applying a thin coat, 9 parts thinner to 1 part paint

93 'It will occur when the surface temperature of a newly applied paint film is lowered by solvent evaporation sufficient to precipitate atmospheric moisture which may be present.' The foregoing is a description of conditions where the following paint defect could well occur:
   (a) blushing or blooming
   (b) orange peel
   (c) dry spray
   (d) craters

94 Correct material viscosity is emphasized to obtain the best results, yet a mix in the proportions of 9 parts thinner to 1 part of paint is recommended for
   (a) applying etch primer
   (b) a guide coat
   (c) spraying metallic finishes
   (d) final colour coat

95 Which of the following defects will not necessarily require repainting for rectification?
   (a) water spotting
   (b) thin paint
   (c) blisters
   (d) off-colour

96 When using a metallic nitrocellulose lacquer, the air pressure is adjusted to 3.15 bar (45 psi), the material adjusted to 35 seconds viscosity, and the spray gun adjusted to a narrow fan. If full wet coats are then applied it will tend to produce a finish with
   (a) a silvery or metallic face tone
   (b) a darker, more colourful face tone with a granular metallic effect
   (c) a normal metallic appearance
   (d) a very fine metallic appearance

97 Some open-coat abrasive papers available for dry flatting contain stearate powder as a dry lubricant. There is a precaution to observe if this type of abrasive paper is used.
   (a) it should not be used to flat colour coats since the stearate powder can cause straining of the lighter colours
   (b) it should not be used to flat primer–

surfacer coats since these are relatively porous and some stearate powder could be retained to give rise to poor adhesion of the colour coats
(c) it should not be used on orbital power sanders
(d) it should not be used to flat acrylic paint

98 When a combination of (a) high material viscosity, (b) low air pressure, (c) hot dry spraying conditions occurs, which of the following paint defects is most likely to result?
(a) colour sinkage
(b) pits
(c) low gloss
(d) water spotting

99 For safety reasons cellulose paint should be stored
(a) in the general stores
(b) in separate purpose-built stores
(c) on the paint mixing bench
(d) on shelves in the workshops

100 Before turning on compressed air it is essential to check the
(a) compressor switch
(b) air receiver
(c) airline connections
(d) water traps

Answers to multiple-choice questions

1	c	26	b	51	b	76	b
2	c	27	c	52	c	77	b
3	b	28	b	53	a	78	a
4	a	29	b	54	a	79	d
5	c	30	c	55	d	80	b
6	b	31	b	56	b	81	c
7	b	32	a	57	b	82	c
8	c	33	c	58	b	83	d
9	d	34	b	59	b	84	c
10	b	35	b	60	a	85	c
11	b	36	a	61	c	86	c
12	d	37	b	62	d	87	a
13	b	38	c	63	a	88	c
14	b	39	b	64	d	89	d
15	b	40	c	65	c	90	a
16	a	41	b	66	d	91	c
17	b	42	b	67	b	92	c
18	a	43	b	68	a	93	a
19	a	44	c	69	c	94	b
20	b	45	a	70	c	95	a
21	c	46	d	71	c	96	b
22	b	47	b	72	b	97	b
23	d	48	a	73	a	98	b
24	b	49	d	74	a	99	b
25	c	50	c	75	b	100	c

Index

acetylene, 54
acid: oxalic, 153; phosphoric, 153
acrylic: colour system, 161; lacquers, 143; modified synthetic, 163; thermoplastic, 158; thermosetting plastic, 158
Act: Deposit of Poisonous Waste, 34; The Control of Pollution, 34; The Factories, 35; The Health and Safety at Work etc., 33; The Petroleum Consolidation, 33
adhesion, 139, 170
administration, 104
air: cap, 118; compressor, 105; hoses, 107; regulators/filters, 107; tool oilers, 107; tools, 24
alignment: chassis, 39; jigs, 18; wheel, 39
aluminium, 61; brazing, 61; particles, 147
anchorage, 89, 91
angling, 128
anti-sag, 139
arc welding, metallic, 75
atomization, 126

banding, 180
bead, 45
beaters, 40; shrinking, 49
beating, 40; direct, 42; indirect, 42; spring, 42
bench: dedicated, 95; repair, 94; universal, 95
binder, 141
bleeding, 139, 170
blistering, 171
blocking, 44
blushing, 172
body repair shop, requirements of, 17
booth: open-ended, 114; oven, combined spray, 114; peelable coating, 114
brass, 63
British Standards Institution, 31
bronz, 63
bronzing, 172
buckling, 39
bumping blades, 48

carbon monoxide, 55
cast iron, 61

cellulose: colour system, 158; lacquer, 158
chalking, 172
chemical resistance, 139
chipping, 173
clamping, 86
cleaning: equipment, 130; gravity feed gun, 130; pressure feed system, 130; suction feed gun, 129
cleanliness, 38
clouding, 139
coachlines, narrow, 156
coating, 139
coats: double, 135; single, 135
colour, 139, 173; coat, 151; tone, 139
compressed air, 29; blow guns, 123
compressor, 105
consumption, 139
contact: eye, 38; skin, 37
copper, 63
corrosion, protection against, 139
coverage, effective, 140
cracking, 174
crazing, 146, 174
cross-spraying, 149
crown, 10
crowned panels, shaping of, 44

damage: assessment, 39; direct, 39; indirect, 39
danger sign, 33
decor, self-adhesive, 156
decoration, 104
de-waxing, 154; fluid, 154
diamond, 88, 89
dinging, 42
dirt, 175
dolly, 40; long-reach, 48; on-and-off the, 42; shrinking, 50; grid, 49
double-insulated tools, 25
Dozer, 81; Damage, 82; Pull, 82; set-ups, 87; technique, 81; Unit, 82
dry application, 147; coats, 181; spray, 175
drying, 159, 176; lamps, infra-red, 119; methods, 139; process, 141; stages, 139; time, 139
dust extraction, 104
dusting, technique, 149

elasticity, 39
electric tools, 25
electricity, 29; static, 35
electrostatic spraying, 137; principles of, 137
employees, 34
employers, 33
enamels: air-dry synthetic, 144; low-bake, 144; two-pack acrylic, 144
estimate, 15
extinguishers, 35

fading, 176
feather edge, 152
filing, 51
fillers, 142
filling, 52; capacity, 140
finishes, 143
fire extinguisher, types of, 35
fish-eyes, 177
flange, 44
flashback, 55
flash-off periods, 135
flatting scratches, 177
flip, 148
flooding, 178
floors, 104
flow cup, 124
flow structure, 140
formulation, 147
frame gauges, 17

gasses: care and handling of, 55; explosions, 55; safe handling, 55; storage of, 55
glass-fibre reinforced plastics (GFRP) bodywork, 29
gloss, 178; retention, 140
gravity feed, 115
grinder: bench, 27; straight, 26
grinderette, 26
ground coats, 143

hardness, 140
hazing, 140
heater, paint, 136
health and safety, 28
hiding, 179
hold-out, 179
hollowing, 43
hot spray cup, 119
hydraulic, 78
hydrogen, 55

ignition, sources of, 34
inhibitor sealers, 143
insurance, 10
isocyanate ingestion, 38
isocyanate paints, 35
isolators, 143

jack body, 40
jack floor, 96
jig, 39
jig saw, 26

kick-up, 88, 89
Kitemark, British Standard, 25
Korek repair system, 92

lacquers, 145
lifting, 180
lighting, 104

main air lines, 107
malleable iron, 61
mallets: bossing, 43; copper-faced, 43; pear-shaped, 43; rubber, 43; standard round, 43
metal arc gas-shielded (MAGS) welding, 69; advantages, 72; arc and power polarity, 69; definition of, 69; dip transfer, 71; metal transfer, 70; safety precautions, 72; spray transfer, 71; torches, types of, 71; types, 69
metallescent, 146
metallic appearance, 148; clear-coat, 150, 168; colours, 159; effect, 148; finishes, 146; single-coat, 148
methyl, methacrylate, 143
methylated spirits, 129
mixing room, 109
mixing scheme, 109
motion study, 133

naphtha, 150
needle, fluid, 119
nibbling, 25
Nike Dataliner, 99
nitrocellulose, 158; lacquers, 143; synthetic, 158

opacity, 150
opalescent, 146
orange peel, 181
overlap, 129
oxy-acetylene welding, 54; blowpipe, 57; cylinders,

57; edge preparations, 60; equipment, 56; flame conditions, 59; gas economizer, 57; leftward, 59; lighting up procedure, 58; nozzles, 57; of various metals, 60; regulators, 57; rightward, 60; shutting-down procedure, 59; vertical, 60
oxygen, 54; cutting, 64

paint: characteristics, 141; coat, 140; film, 140; foundation, 140; heating, 135; material, 140; store, 105; substance, 140
paint film thickness gun, 122
panel beating, 39; small blemish, 150
panel finishing, 49
panel puller, 48
patterns: defective spray, 126; spray, 125
peaks, 39
penetration, 140
personnel, 38
pigment, 141
pinholing, 182
plan, spraying, 134
polish, 150
polishable, 140
polisher, high-speed, 122
polychromatic, 146
portable power tools, 24
Porto-Power, 78, 83
precipitation, 182
preparation, 108; panel, 40; surface, 160
pressing, 87
pressure, atomizing, 126; feed, 115
pressure regulator gun, 119
primer, 140; coat, 140; fillers, 142; material, 140; surfacers, 142
primers, 142, 160; etch, 161; synthetic/alkyd, 165; synthetic resin, 160; /undercoats, 167
procedure: spraying, 133; suggested, 134
protective clothing, 37
puckers, 44
pulling, 84
pull-posts, 91
pushing, 83

quick-release couplings, 107

raising, 43
receptacle, gun cleaning, 109
refinish processes, 150
relationship, gun surface, 128
remote cup, 116
respirators, 114

respray, local, 160
ridges, 39
ringing, 181
rules, basic refinishing, 123
runs, 140, 182
rusting, 183

safety, 27
sag, 88
sagging, 140
sander: disc, 27; finishing, 27
sanders, 120
sanding ability, 140
scratch resistance, 140
sealer, 143
sedimentation, 140
separation, 140
settling, 183
shrinking, 49; hot, 50
signs: mandatory, 31; prohibition, 31; safe condition, 33; safety, 31; supplementary, 33; warning, 31
skinning, 184
smoking, 35
solvent, 141
spillage, 38
spoons, 46; general-purpose, 46; heavy-duty pry, 47; high-crown, 47; inside pry, 46; spring-beating, 47; surfacing, 46
spot repairs, 154
spot welding, 66; hold time, 68; series or twin, 68; squeeze time, 67; the principles of, 66; weld time, 68
spray booth, 109
spray gun, 114; adjustments, 125; construction, 117; lubrication, 132; maintenance, 129; problems, 132; set-up, 117; small capacity, 119
spreading, 85
staining, 184
stainless steel, 61
starfire, 146
starmist, 146
steam cleaning, 154
stoppers, 143, 161
streaking, 185
strength of metal, 9
stretching, 85
sway, 88
swelling, 140
synthetic colour system, 166; low-bake, 167

technique, spraying, 127
tensioning, 85

test drop, 17
test spray pattern, 126
thinner, 141, 146
thinning, 159
tidiness, 38
tip, fluid, 118
tone, 148; bottom, 148; mass, 148; top, 148; under, 148
tools, care of, 52
toxic effects, 37
treatment, emergency, 38
triggering, 126
tungsten arc gas-shielded (TAGS) welding, 73; arc starting, 74; characteristics, 75; current, 75; electrodes, 74; electrode negative, 74; electrode polarity, 74; gas shield, 74; principle of operation, 73
twist, 88, 89

undercoats, 142
undertone, 148
unit construction, 15

vehicle, 141
velocity, air, 104
ventilation, 104
viscosity, 122, 124, 146

water: leak diagnosis, 155; spotting, 185; supply and drainage, 103
weather resistance, 141
wet application, 147
wetting, 141; interference, 141
work hardened, 9
wrinkling, 186

yellowing, 141